FOURIER SERIES and HARMONIC ANALYSIS

K A STROUD

Formerly Principal Lecturer, Department of Mathematics,
Lanchester Polytechnic, Coventry

Stanley Thornes (Publishers) Ltd

First published 1984 by

Stanley Thornes (Publishers) Ltd
Educa House
Old Station Drive
Leckhampton
CHELTENHAM GL53 0DN

British Library Cataloguing in Publication Data

Stroud, K A
Fourier series and harmonic analysis.
1. Fourier series
I. Title
515'.2433 QA404

ISBN 0-85950-137-X

Typeset in 11/12 IBM Press Roman by Tech-Set, Gateshead, Tyne & Wear.
Printed and bound in Great Britain by Bell and Bain, Thornliebank, Glasgow.

Fourier Series and Harmonic Analysis

Also by K A Stroud

Laplace Transforms – Programmes and Problems (Stanley Thornes (Publishers) Ltd)
Engineering Mathematics (Macmillan Ltd)

CONTENTS

Preface vii

Chapter 1: Periodic Functions 1
Graphs of periodic functions
Harmonics
Compound waveforms
Non-sinusoidal periodic functions
Analytical description of a periodic function
Sketching graphs of periodic functions
Revision summary

Chapter 2: Useful Revision 14
Trigonometrical identities
Hyperbolic functions and identities
Integration by parts

Chapter 3: Fourier Series for Functions of Period 2π 23
Periodic functions
General form of Fourier series
Dirichlet conditions
Useful integrals
Fourier coefficients
Convergence of a Fourier series
Fourier series at a discontinuity
Odd and even functions and their products
Fourier series for odd and even functions
Revision summary

Chapter 4: Half-range Series 53
Functions defined over half a period
Half-range sine and cosine series
Series containing only odd harmonics or only even harmonics
Significance of the constant term $\frac{1}{2}a_0$
Revision summary

Chapter 5: Functions with Periods other than 2π 67
Functions with period $2L$
Change of units and alternative method
Functions with period T
Fourier coefficients
Half-range sine and cosine series
Revision summary

Chapter 6: Numerical Harmonic Analysis 83
Approximate integration – trapezoidal rule
Twelve-point analysis
Display in tabular form
Percentage harmonics
Revision summary

Chapter 7: Applications of Fourier Series 104
Half-wave rectifier output
Full-wave rectifier output
Parseval's theorem
RMS value of a periodic voltage or current expressed as a Fourier series
Multiplication theorem
Average power in a circuit
Spectrum of a waveform
Revision summary

Chapter 8: Further Techniques 119
Integration and differentiation of a Fourier series
Gibbs' phenomenon
Fourier coefficients from function value jumps at discontinuities
Complex form of Fourier series
Equivalence of complex and trigonometrical forms
Revision summary

Chapter 9: Solution of Boundary Value Problems 146
Ordinary second order differential equations (summary)
Partial differential equations
The wave equation
Solution by separation of variables
The heat conduction equation for a uniform finite bar
Laplace's equation
Revision summary

Chapter 10: Fourier Integrals and Transforms 169
Extended period
Fourier integral
Other forms of the Fourier integral
Fourier sine and cosine integrals
Amplitude and phase – spectrum of a waveform
Fourier transforms – complex, sine and cosine transforms
Inverse transforms
Fourier integrals in the solution of boundary value problems
Revision summary

Answers 192

Index 203

PREFACE

A knowledge of Fourier series and of the techniques of Fourier analysis is now regarded as of utmost importance in the study of many branches of science and technology, for it provides a powerful tool in the solution of problems of a periodic nature occurring in situations involving electrical and mechanical vibrations, propagation of electromagnetic waves, acoustics, heat conduction, physical fields and the like.

The discovery by Jean Baptiste Joseph Fourier (1768–1830) that all periodic functions can be expressed as a sum of sinusoidal components permits two interpretations of the series representation of such a function:

(a) as an approximation to the function, where normally the first few terms of the infinite series suffice for practical purposes,

(b) as an analysis of the function into its harmonic components.

Both aspects are emphasised in the early chapters of the book.

An important advantage of the Fourier series representation of a function over the corresponding Taylor series representation is that the former can equally well represent a periodic function which contains a number of finite discontinuities, whereas Taylor series requires the use of successive differential coefficients.

Fourier series is a topic appearing in numerous course syllabuses. In some, the subject is pursued as far as the numerical harmonic analysis of a given waveform. This has wide applications and the subject is accordingly developed in detail in the first six chapters, which include also treatment of sine and cosine series representation of odd and even functions, and half-range series.

For those students wishing to proceed to rather wider aspects of the subject, Chapters 7 to 10 cover further applications and techniques, leading to the use of Fourier series in the solution of boundary value problems, Fourier integrals and an introduction to Fourier transforms.

Full mathematical rigour has not been attempted in a book of this size, but sufficient proofs have been included to establish the relevant results and to provide a foundation for the techniques used. Throughout the text, numerous worked examples are provided at each stage, together with sets of graded exercises by which the necessary practice can be undertaken and confidence with the methods assured. A complete set of answers is provided at the back of the book.

The author wishes to record his sincere thanks to all those who have shown an interest in the work and who have offered constructive comment; to acknowledge the many sources from which examples have been gleaned over the years; and to thank the publishers for their valuable advice in the preparation of the text for publication.

K.A.S.

Chapter 1

PERIODIC FUNCTIONS

1.1 GRAPHS OF PERIODIC FUNCTIONS

1.1.1 Characteristics

A *periodic function* is one in which the whole set of function values repeats at regular intervals of the independent variable. A common example of such a function is $\sin\theta$, which goes through its complete range of values as the angle θ increases from 0° to 360°. As θ continues past 360°, all the values of the function repeat, giving the characteristic waveform to the graph of $y = \sin\theta$.

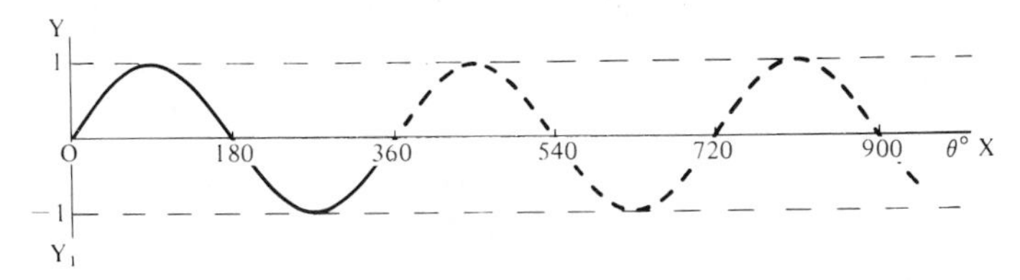

The constant interval of θ, after which repetition occurs, is called the *period* of the function and the part of the waveform extending over one period is referred to as *one cycle*.

The graph of $y = \cos\theta$ also illustrates a periodic function.

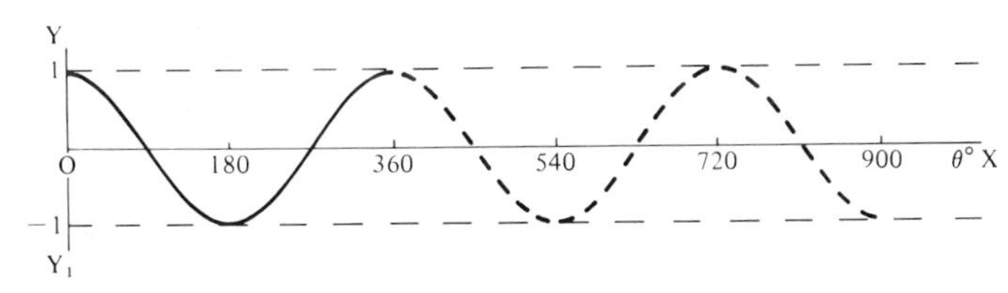

1.1.2 Graphs of $y = A\sin n\theta$

The function $y = A\sin 2\theta$ assumes all its values as the angle 2θ increases from 0° to 360°, i.e. as θ increases from 0° to 180°. Thus, one complete cycle occurs between $\theta = 0°$ and $\theta = 180°$.

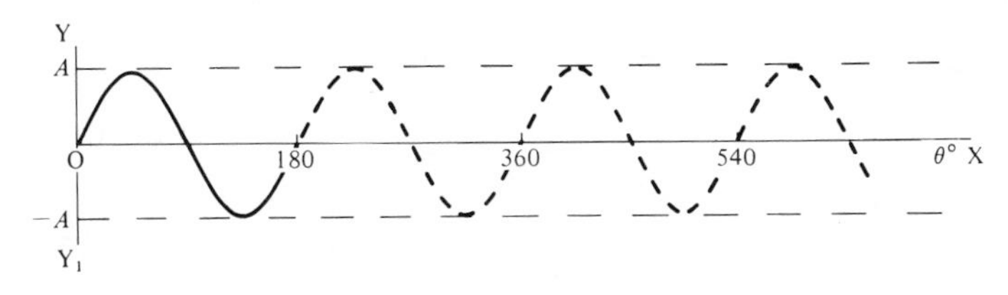

The factor A denotes the *amplitude* of the function, i.e. the maximum displacement of the curve from its mean value.

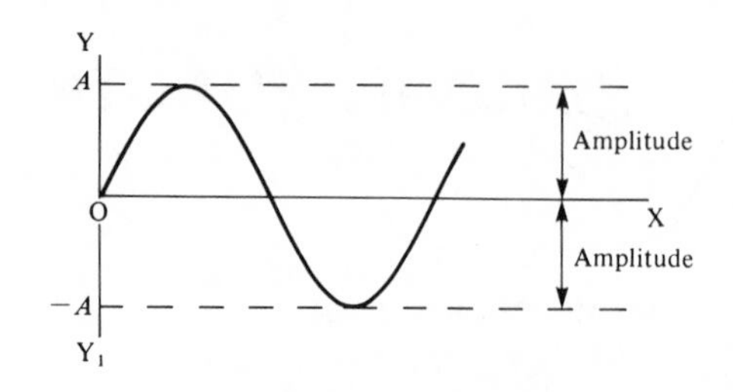

Therefore, for $y = A \sin 2\theta$, the period is given by $2\theta = 360°$, i.e. $\theta = 180°$.
Similarly, for $y = A \sin 3\theta$, the period is given by $3\theta = 360°$, i.e. $\theta = 120°$.
and for $y = A \sin 4\theta$, the period is given by $4\theta = 360°$, i.e. $\theta = 90°$, etc.
In general, for $y = A \sin n\theta$, the period is given by $n\theta = 360°$, i.e. $\theta = \dfrac{360°}{n}$.

Exercise 1

In each of the following cases, sketch the graph of the function over two cycles, indicating (i) the value of the amplitude and (ii) the values of θ at which the graph crosses the axis of θ.

1. $y = 3 \sin \theta$
2. $y = 2 \sin 2\theta$
3. $y = \sin 4\theta$
4. $y = 5 \cos 3\theta$
5. $y = 2 \cos \theta$
6. $y = 4 \sin 5\theta$
7. $y = 2 \sin \dfrac{\theta}{2}$
8. $y = 3 \sin 6\theta$
9. $y = \cos 2\theta$
10. $y = 5 \cos 0.5\theta$

1.2 HARMONICS

1.2.1 Harmonics of $y = A \sin \theta$

Consider again the graphs of $y = A_n \sin n\theta$, where n is a positive integer.

(a) $n = 1$, i.e. $y = A_1 \sin \theta$

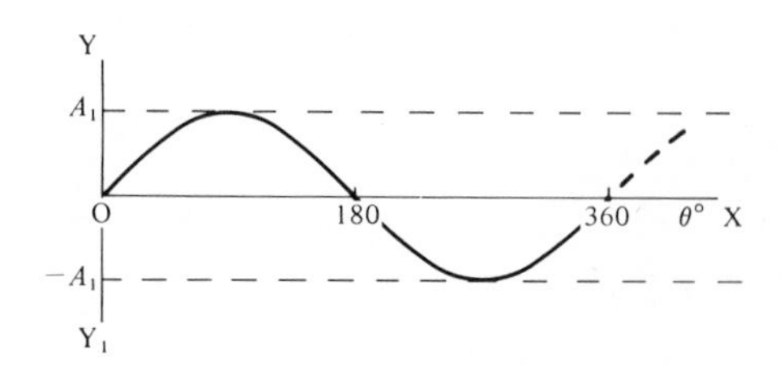

Period = 360°.
1 complete cycle in 360°.
This is the *fundamental* or *first harmonic* of $y = A \sin \theta$.

(b) $n = 2$, i.e. $y = A_2 \sin 2\theta$

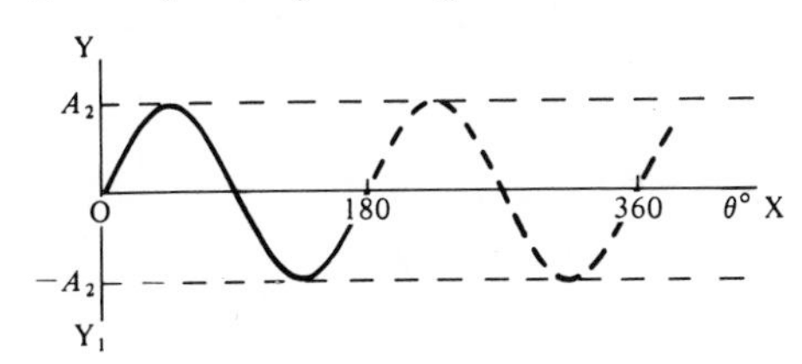

$$\text{Period} = \frac{360°}{2} = 180°.$$

2 complete cycles in 360°.

This is the *second harmonic* of $y = A \sin \theta$.

(c) $n = 3$, i.e. $y = A_3 \sin 3\theta$

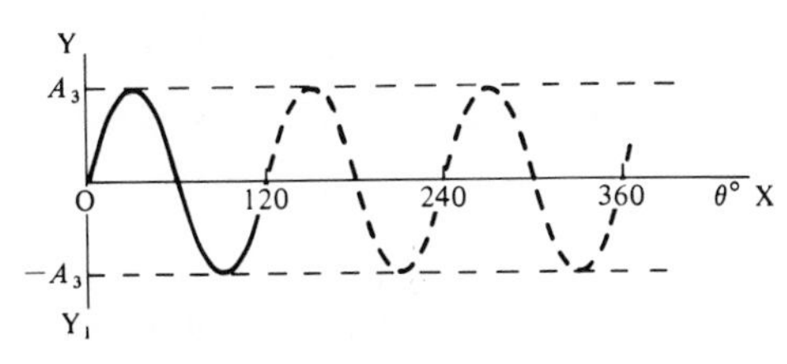

$$\text{Period} = \frac{360°}{3} = 120°.$$

3 complete cycles in 360°.

This is the *third harmonic* of $y = A \sin \theta$.

Therefore, we have

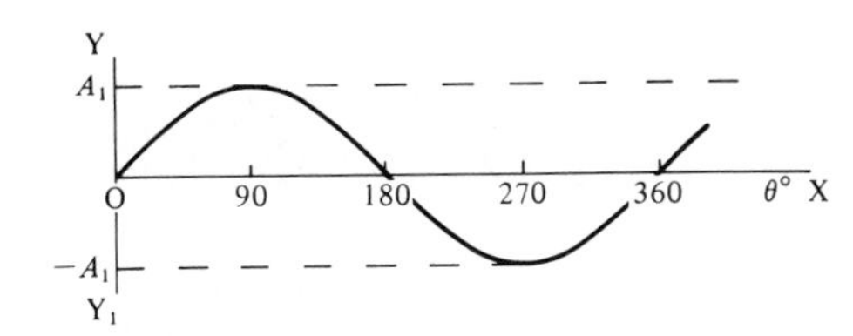

$y = A_1 \sin \theta$

Fundamental or first harmonic.

Period = 360°.

1 cycle in 360°.

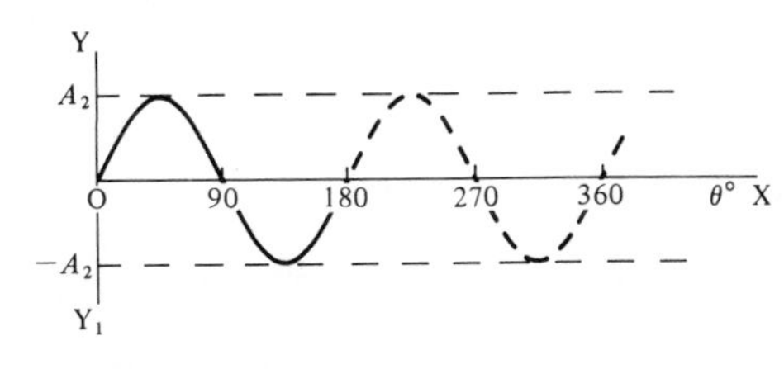

$y = A_2 \sin 2\theta$

Second harmonic.

$$\text{Period} = \frac{360°}{2} = 180°.$$

2 cycles in 360°.

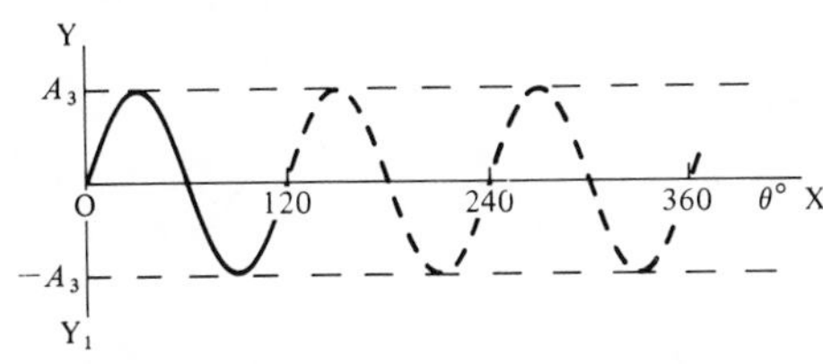

$y = A_3 \sin 3\theta$

Third harmonic.

$$\text{Period} = \frac{360°}{3} = 120°.$$

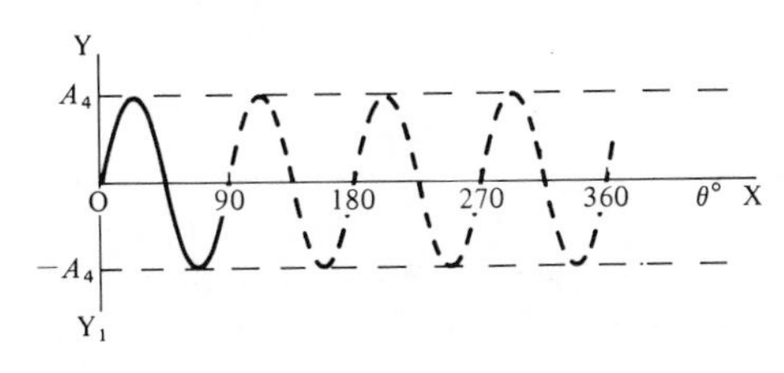

$y = A_4 \sin 4\theta$

Fourth harmonic.

$$\text{Period} = \frac{360°}{4} = 90°.$$

4 cycles in 360°.

In general, the graph of $y = A_n \sin n\theta$ represents the nth *harmonic* of $y = A \sin\theta$. The period $= \dfrac{360^\circ}{n}$ and there are thus n complete cycles in 360° or 2π radians.

1.2.2 Odd and even harmonics

$\sin 2\theta, \sin 4\theta, \sin 6\theta, \ldots,$ are *even* harmonics of $\sin\theta$.

$\sin 3\theta, \sin 5\theta, \sin 7\theta, \ldots,$ are *odd* harmonics of $\sin\theta$.

So, with n an even integer, $y = A_n \sin n\theta$ indicates even harmonics and with n an odd integer, $y = A_n \sin n\theta$ indicates odd harmonics.

1.2.3 Graphs of $y = A_n \cos n\theta$

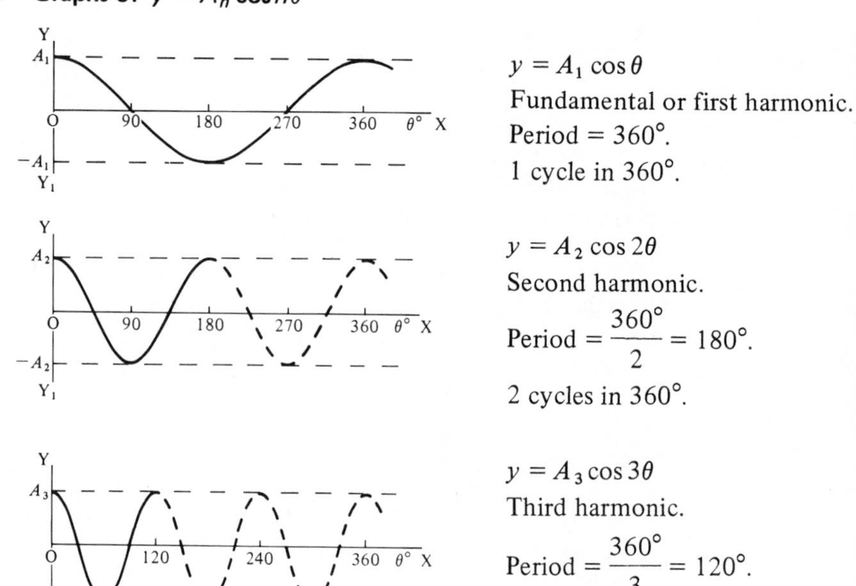

In general, the graph of $y = A_n \cos n\theta$ represents the nth harmonic of $y = A \cos\theta$. The period $= \dfrac{360^\circ}{n}$ and there are thus n complete cycles in 360° or 2π radians.

1.3 COMPOUND WAVEFORMS

Compound waveforms can result from the addition of two or more sine or cosine curves, often a fundamental and one or more harmonics.

Example 1

To obtain the graph of $y = 3\sin x + \sin 2x$, we can plot the graphs of $y = 3\sin x$ and $y = \sin 2x$ separately and add the ordinates.

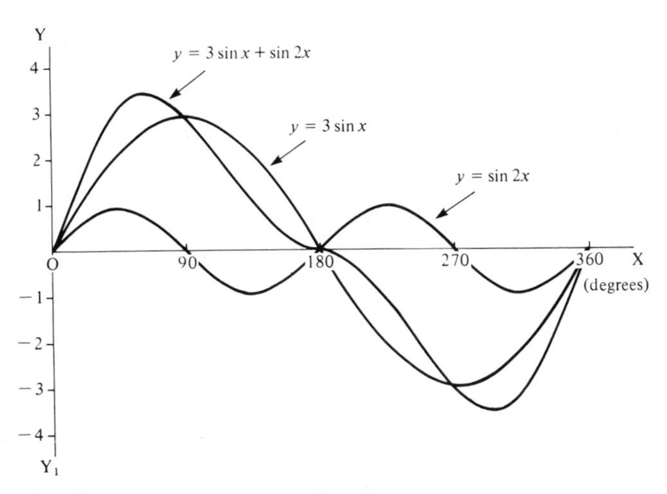

Note that the period of $y = 3 \sin x$ is 360° and that of $y = \sin 2x$ is 180°.

Therefore, there are two cycles of $y = \sin 2x$ in the one cycle of $y = 3 \sin x$.

The period of the combined function, $y = 3 \sin x + \sin 2x$, is 360°, which is the period of its fundamental component.

Example 2

To obtain the graph of $y = 4 \sin 2x + \sin 6x$, we proceed very much as before.

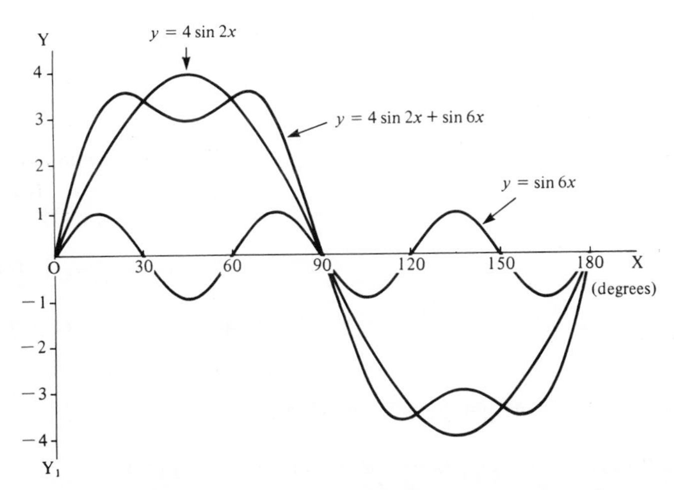

In this case, $y = \sin 6x$ is the third harmonic of $y = 4 \sin 2x$.

The period of $y = 4 \sin 2x$ is 180°; the period of $y = \sin 6x$ is 60°.

Therefore, there are three complete cycles of $y = \sin 6x$ in one complete cycle of $y = 4 \sin 2x$.

The period of $y = 4 \sin 2x + \sin 6x$ is therefore 180°.

Example 3

To obtain the graph of $y = 2 \sin x + \cos 3x$.

Before we plot the curves, we know that the period of $y = 2 \sin x$ is $360°$ and that the period of $y = \cos 3x$ is $120°$. Therefore there are three complete cycles of $y = \cos 3x$ in one complete cycle of $y = 2 \sin x$.

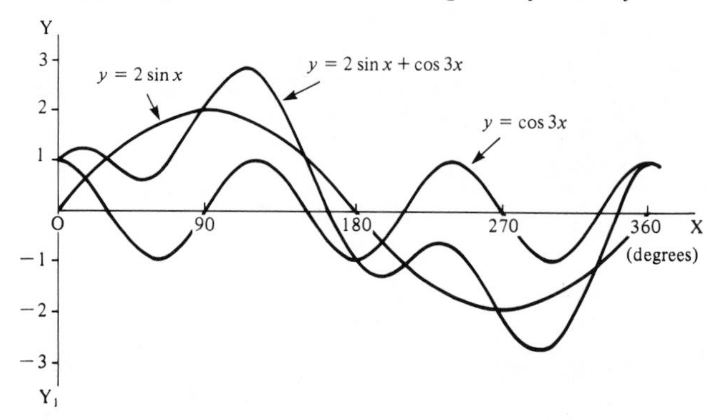

Many technological situations give rise to compound periodic waveforms and, in order to study one such output, it is convenient to analyse it into a number of component sine and cosine constituents. Such a process is therefore the reverse of that employed in the three examples worked above. Just how this analysis is achieved is explained in detail in a later chapter.

Exercise 2

Without drawing any graphs, state the period of each of the following functions.

1. $y = 4 \sin 2x$
2. $y = 2 \cos 5x$
3. $y = 3 \sin \dfrac{x}{2}$
4. $y = \sin 4x$
5. $y = 5 \cos 3x$
6. $y = 2 \sin \dfrac{3x}{5}$
7. $y = 5 \sin x$
8. $y = 4 \sin x + 3 \sin 2x$
9. $y = 2 \sin x + \cos 3x$
10. $y = 6 \sin 2x + 2 \sin 4x$

1.4 NON-SINUSOIDAL PERIODIC FUNCTIONS

Not all periodic functions are sinusoidal in appearance.

Examples

1. Y

O 2 4 6 8 10 X

Period t (ms)

Period = 4 ms

2.

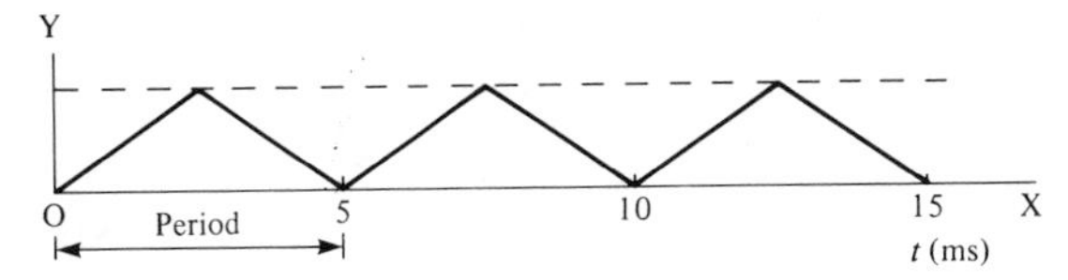

Period = 5 ms

3.

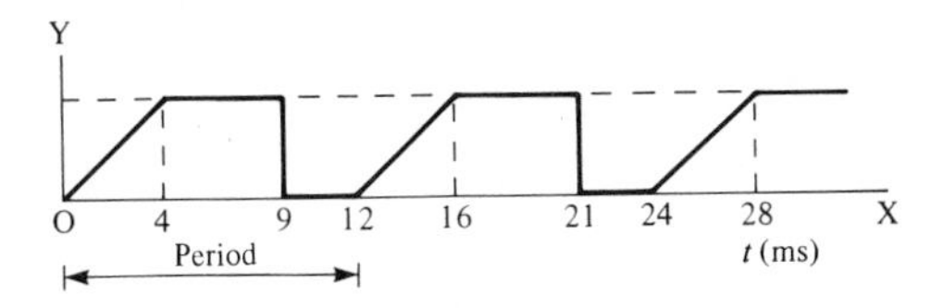

Period = 12 ms

4.

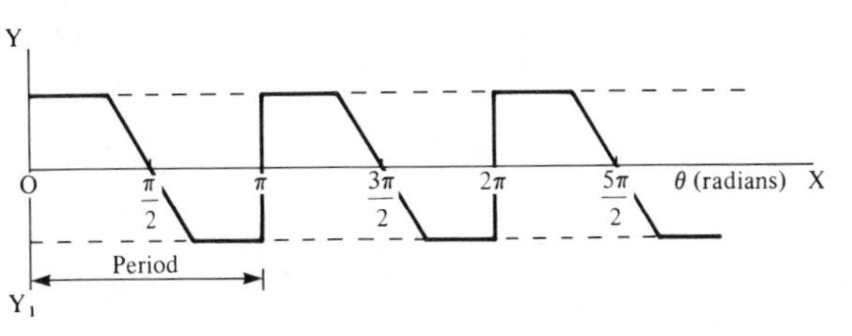

Period = π radians

In each case, the period is the shortest interval of the independent variable before repetition occurs.

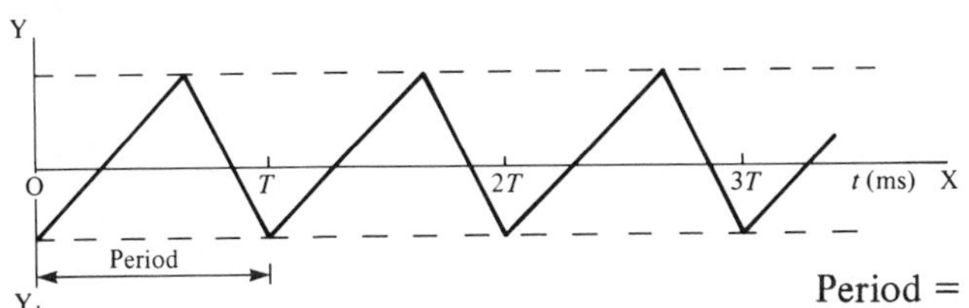

Period = T milliseconds

Exercise 3

In each of the following cases, the independent variable is time in milliseconds. Write down the period of each waveform shown.

1.

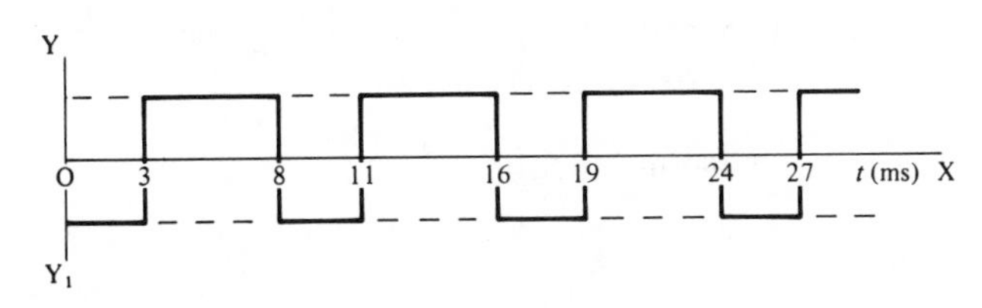

2.

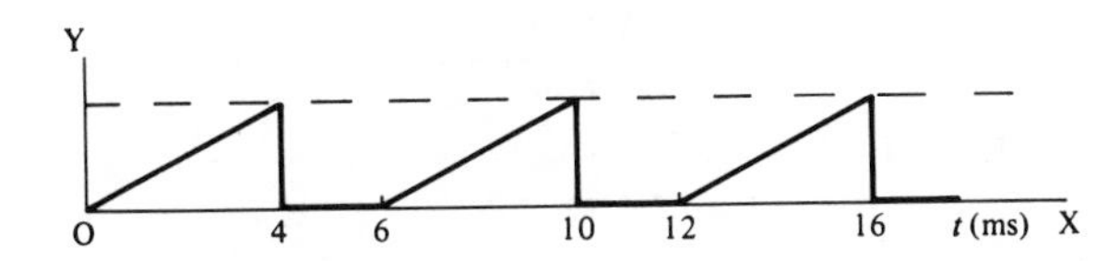

3.

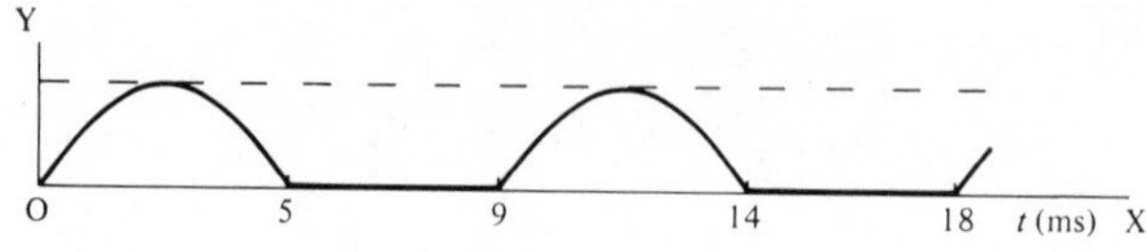

4.

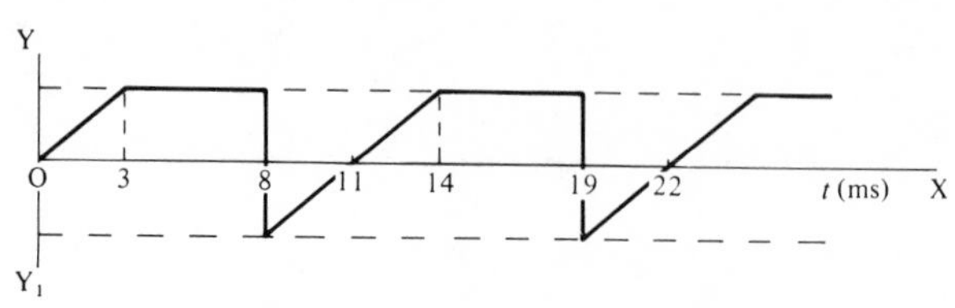

5.

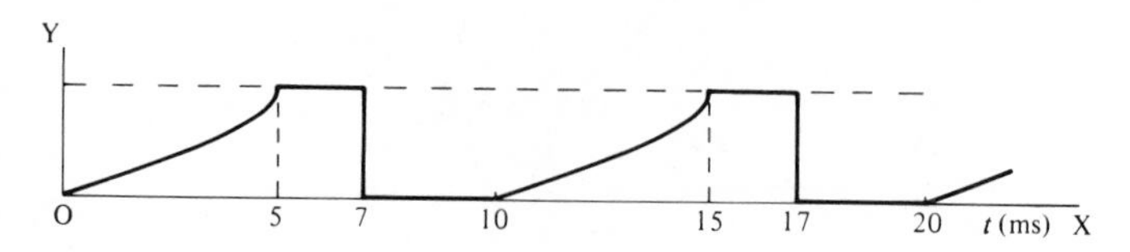

1.5 ANALYTICAL DESCRIPTION OF A PERIODIC FUNCTION

1.5.1 Given waveforms

It is often convenient to describe a periodic function algebraically.

Example 1

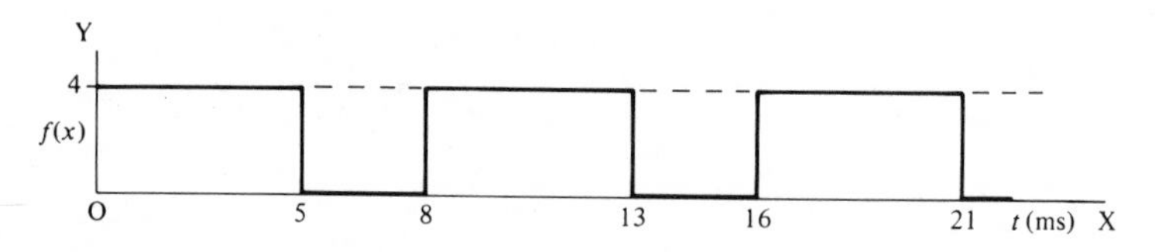

We see that

(a) between $x = 0$ and $x = 5$, the graph is the line $y = 4$. This can be written

$$f(x) = 4 \qquad 0 < x < 5$$

(b) between $x = 5$ and $x = 8$, the graph is the line $y = 0$,

i.e. $$f(x) = 0 \qquad 5 < x < 8$$

(c) the period of the function is 8 units.

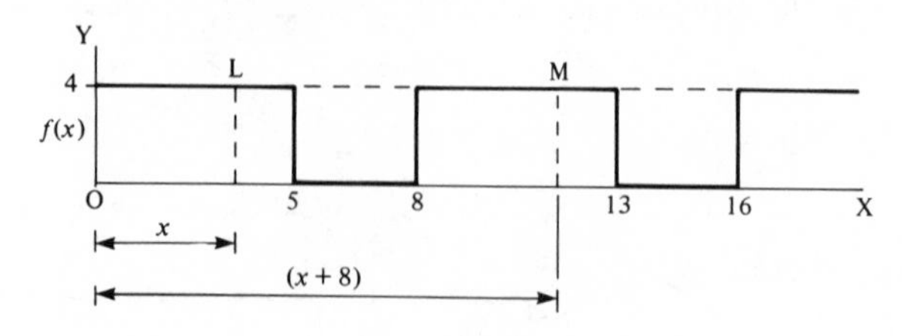

Therefore, we can completely define the function in the form

$$f(x) = 4 \qquad 0 < x < 5$$
$$f(x) = 0 \qquad 5 < x < 8$$
$$f(x) = f(x+8)$$

Example 2

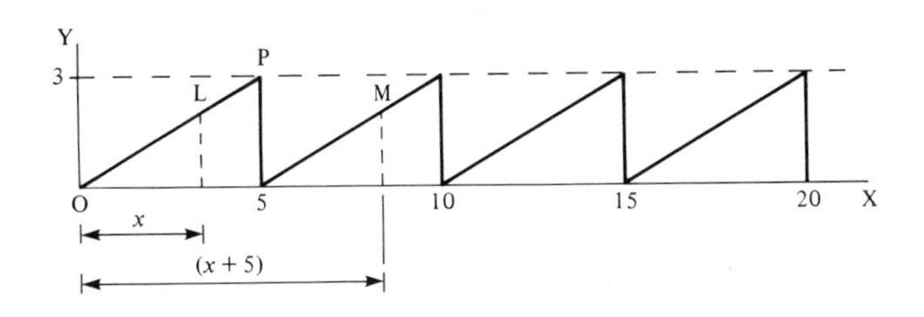

The equation of OP is $y = \frac{3}{5}x$.

(a) Between $x = 0$ and $x = 5$, the graph is the line $y = \frac{3}{5}x$,

i.e. $$f(x) = \tfrac{3}{5}x \qquad 0 < x < 5$$

(b) The period of the function is 5,

i.e. $$f(x) = f(x+5)$$

Therefore, the periodic function can be defined as

$$f(x) = \tfrac{3}{5}x \qquad 0 < x < 5$$
$$f(x) = f(x+5)$$

Example 3

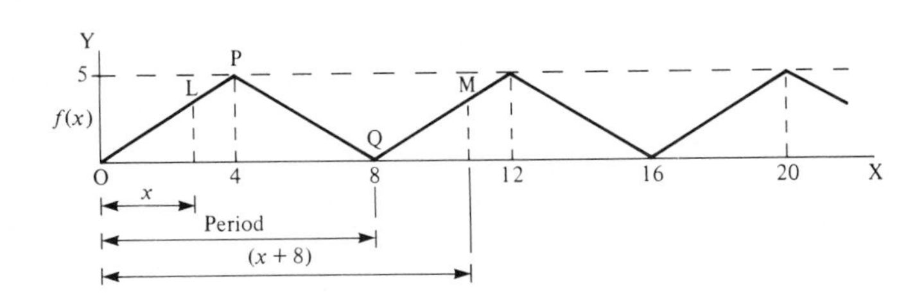

(a) The equation of OP is $y = \frac{5}{4}x$.

$$\therefore \; f(x) = \tfrac{5}{4}x, \qquad 0 < x < 4$$

(b) The equation of PQ is $y = -\frac{5}{4}x + 10$.

$$\therefore \; f(x) = -\tfrac{5}{4}x + 10, \qquad 4 < x < 8$$

(c) Period = 8

$$\therefore \; f(x) = f(x+8)$$

Therefore the periodic function is defined as

$$f(x) = \tfrac{5}{4}x \qquad 0<x<4$$
$$f(x) = -\tfrac{5}{4}x + 10 \qquad 4<x<8$$
$$f(x) = f(x+8)$$

Exercise 4

Define analytically the periodic functions shown.

1.

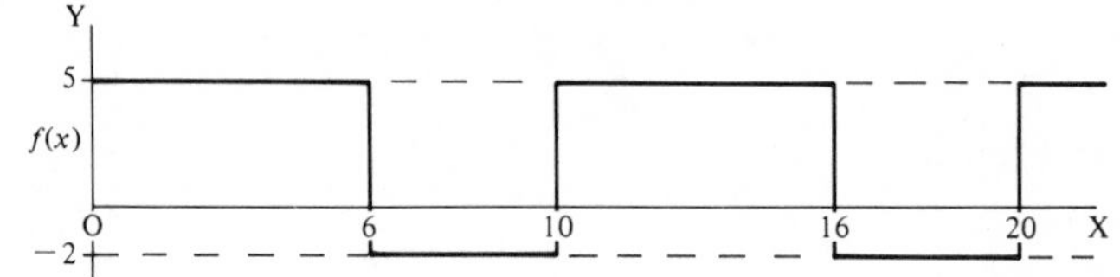

2.

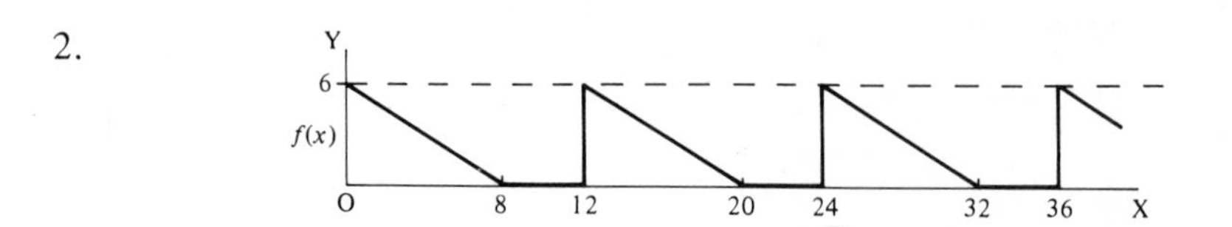

3.

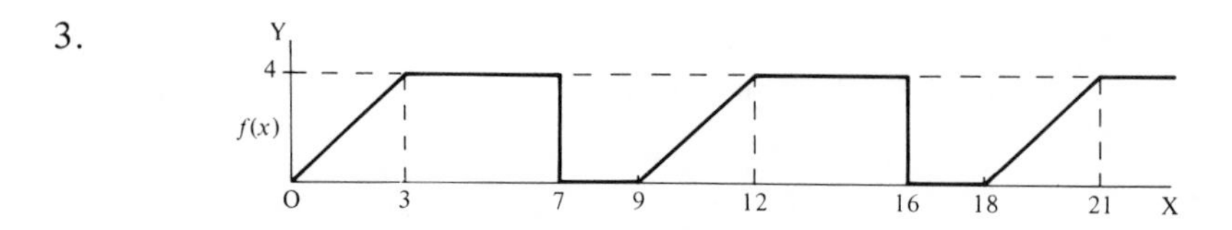

4.

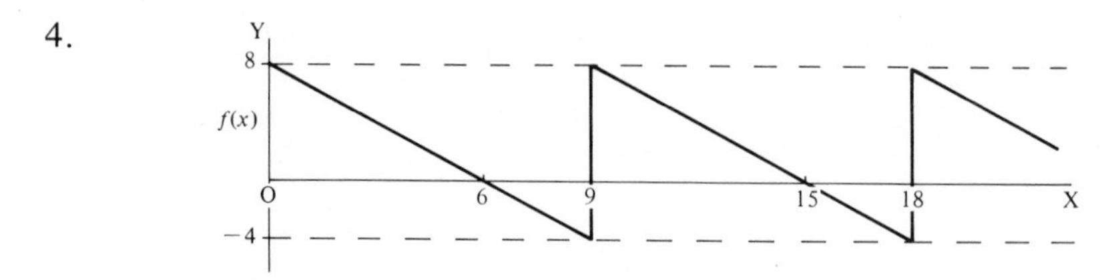

5.

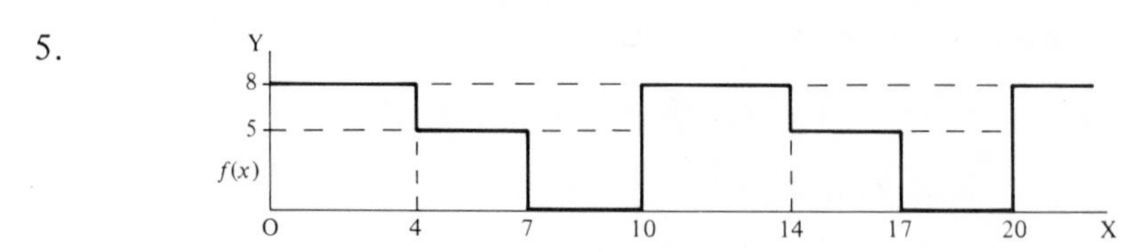

6.

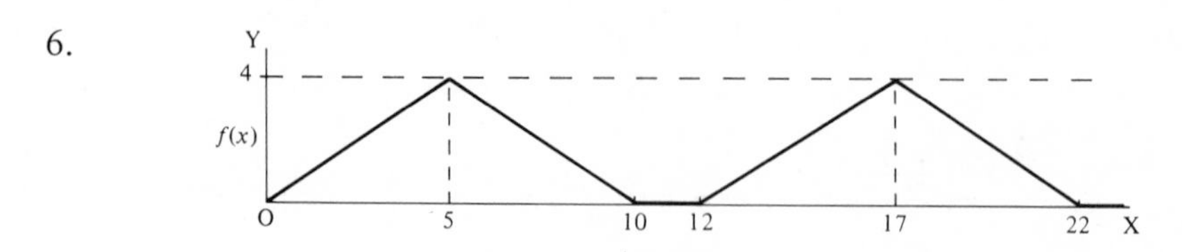

1.5.2 Sketching graphs of periodic functions

If the function is stated in analytical terms, the same considerations can be given to sketch the graph of the function.

Example 1

A function is defined by

$$\begin{aligned} f(x) &= 2x & 0<x<5 \\ f(x) &= -3 & 5<x<8 \\ f(x) &= f(x+8) \end{aligned}$$

Between $x=0$ and $x=5$, the graph is $y=f(x)=2x$.

Between $x=5$ and $x=8$, the graph is $y=f(x)=-3$.

Also $f(x)=f(x+8)$, i.e. any point on the graph is repeated 8 units later, so the period is 8.

Therefore, the graph is

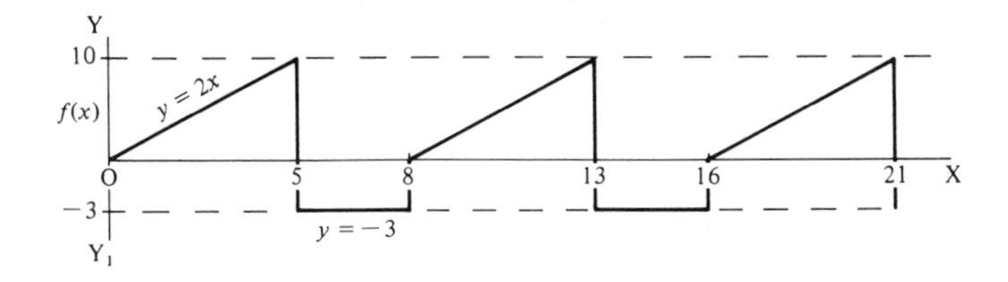

Example 2

Sketch the function defined by

$$\begin{aligned} f(x) &= x^2 & 0<x<4 \\ f(x) &= 16 & 4<x<6 \\ f(x) &= 0 & 6<x<10 \\ f(x) &= f(x+10) \end{aligned}$$

Between $x=0$ and $x=4$, $y=x^2$

x	0	1	2	3	4
y	0	1	4	9	16

So we have

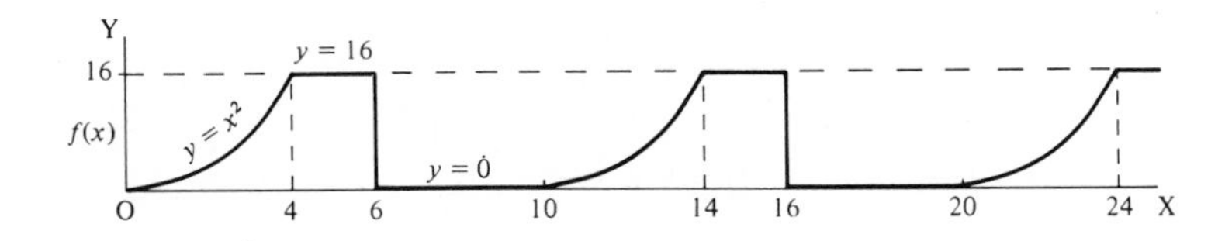

Exercise 5

Sketch the graphs of the following functions.

1. $f(x) = 5 \qquad 0 < x < 3$
   $f(x) = -2 \qquad 3 < x < 7$
   $f(x) = f(x + 7)$

2. $f(x) = -3 \qquad 0 < x < 5$
   $f(x) = 4 \qquad 5 < x < 8$
   $f(x) = f(x + 8)$

3. $f(\theta) = \dfrac{\theta}{2} \qquad 0 < \theta < \pi$
   $f(\theta) = -\dfrac{\theta}{2} + \pi \qquad \pi < \theta < 2\pi$
   $f(\theta) = f(\theta + 2\pi)$

4. $f(x) = -\dfrac{7}{\pi}x + 5 \qquad 0 < x < \pi$
   $f(x) = -2 \qquad \pi < x < \dfrac{3\pi}{2}$
   $f(x) = f\left(x + \dfrac{3\pi}{2}\right)$

5. $f(\theta) = \sin\theta \qquad 0 < \theta < \dfrac{\pi}{2}$
   $f(\theta) = 1 \qquad \dfrac{\pi}{2} < \theta < \pi$
   $f(\theta) = f(\theta + \pi)$

1.6 REVISION SUMMARY

1. Periodic functions

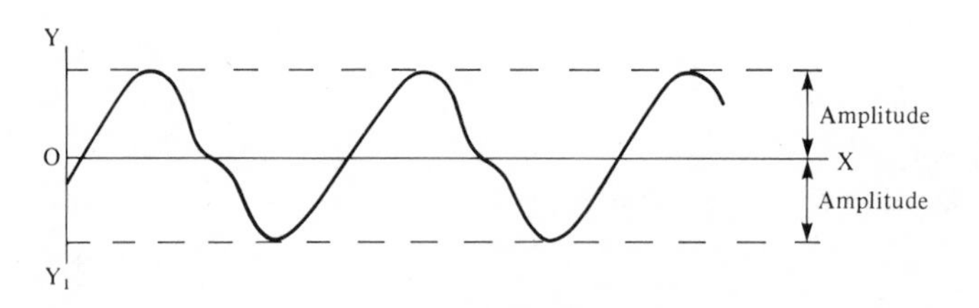

A compound waveform can be expressed as the sum of sinusoidal components.

2. Sinusoidal functions

$$\left.\begin{aligned} y &= A\sin n\theta \\ y &= A\cos n\theta \end{aligned}\right\} \qquad A = \text{amplitude}$$

3. Harmonics of $y = A \sin\theta$

$$\text{Period} = 2\pi = 360°$$

$$y = A_1 \sin\theta = \text{first harmonic (or fundamental)} \quad \text{Period} = \frac{360°}{1} = 360°$$

$$y = A_2 \sin 2\theta = \text{second harmonic} \qquad \text{Period} = \frac{360^\circ}{2} = 180^\circ$$

$\vdots$

$$y = A_n \sin n\theta = n\text{th harmonic} \qquad \text{Period} = \frac{360^\circ}{n}$$

4. Analytical description of a periodic function

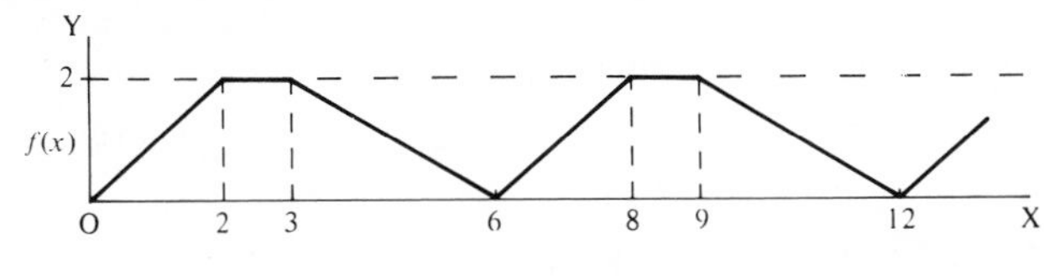

$$f(x) = x \qquad 0 < x < 2$$

$$f(x) = 2 \qquad 2 < x < 3$$

$$f(x) = 4 - \frac{2x}{3} \qquad 3 < x < 6$$

$$f(x) = f(x+6), \quad \text{i.e. period} = 6 \text{ units}$$

Chapter 2

USEFUL REVISION

In the following chapters, a number of essential mathematical techniques will be used repeatedly and it may well be helpful, therefore, to revise these topics before the need arises to apply them in future sections of the work.

2.1 TRIGONOMETRICAL IDENTITIES

2.1.1 Compound angle formulae

$$\sin(A+B) = \sin A \cos B + \cos A \sin B \qquad (2.1)$$

$$\sin(A-B) = \sin A \cos B - \cos A \sin B \qquad (2.2)$$

$$\cos(A+B) = \cos A \cos B - \sin A \sin B \qquad (2.3)$$

$$\cos(A-B) = \cos A \cos B + \sin A \sin B \qquad (2.4)$$

2.1.2 Sum and difference formulae

By addition or subtraction of pairs of the compound angle formulae, we obtain the following results

(2.1) + (2.2) $$2\sin A \cos B = \sin(A+B) + \sin(A-B) \qquad (2.5)$$

(2.1) − (2.2) $$2\cos A \sin B = \sin(A+B) - \sin(A-B) \qquad (2.6)$$

(2.3) + (2.4) $$2\cos A \cos B = \cos(A+B) + \cos(A-B) \qquad (2.7)$$

(2.3) − (2.4) $$-2\sin A \sin B = \cos(A+B) - \cos(A-B) \qquad (2.8)$$

Notice the negative sign on the left hand side of the last result.

If we now write $A+B=C$ and $A-B=D$, then $A=\dfrac{C+D}{2}$ and $B=\dfrac{C-D}{2}$

and results (2.5) to (2.8) become

$$\sin C + \sin D = 2\sin\frac{C+D}{2}\cos\frac{C-D}{2} \qquad (2.9)$$

$$\sin C - \sin D = 2\cos\frac{C+D}{2}\sin\frac{C-D}{2} \qquad (2.10)$$

$$\cos C + \cos D = 2\cos\frac{C+D}{2}\cos\frac{C-D}{2} \tag{2.11}$$

$$\cos D - \cos C = 2\sin\frac{C+D}{2}\sin\frac{C-D}{2} \tag{2.12}$$

In order to keep the pattern of the right hand side consistent, note that we have reversed C and D on the left hand side in the last identity.

2.1.3 Conversion of $a\sin\theta + b\cos\theta$ into the form $r\sin(\theta+\alpha)$

From identity (2.1) above,

$$\begin{aligned} r\sin(\theta+\alpha) &= r\sin\theta\cos\alpha + r\cos\theta\sin\alpha \\ &= (r\cos\alpha)\sin\theta + (r\sin\alpha)\cos\theta \end{aligned}$$

Writing this in reverse and comparing the left hand side with $a\sin\theta + b\cos\theta$

$$(r\cos\alpha)\sin\theta + (r\sin\alpha)\cos\theta = r\sin(\theta+\alpha)$$

$$a \quad \sin\theta + \quad b \quad \cos\theta$$

then $r\cos\alpha = a$ and $r\sin\alpha = b$

$$\therefore\ r^2\cos^2\alpha + r^2\sin^2\alpha = a^2+b^2$$

$$\therefore\ r^2(\cos^2\alpha+\sin^2\alpha) = a^2+b^2 \quad \therefore\ r = \sqrt{a^2+b^2}$$

Also $$\frac{r\sin\alpha}{r\cos\alpha} = \frac{b}{a} \quad \therefore\ \tan\alpha = \frac{b}{a} \quad \therefore\ \alpha = \arctan\frac{b}{a}$$

$$\therefore\ \underline{a\sin\theta + b\cos\theta = r\sin(\theta+\alpha)}$$

where $$r = \sqrt{a^2+b^2}$$

$$\alpha = \arctan\frac{b}{a}$$

In practice, the positive root for r is taken; the auxiliary angle α, or phase, can be in any quadrant depending on the signs of a and b.

Similar results can be obtained by using results (2.2) to (2.4) of the list of identities, so that, in all, we have

$$a\sin\theta + b\cos\theta = r\sin(\theta+\alpha) \tag{2.13}$$

$$a\sin\theta - b\cos\theta = r\sin(\theta-\alpha) \tag{2.14}$$

$$a\cos\theta - b\sin\theta = r\cos(\theta+\alpha) \tag{2.15}$$

$$a\cos\theta + b\sin\theta = r\cos(\theta-\alpha) \tag{2.16}$$

In each case, $r = \sqrt{a^2+b^2}$ and $\alpha = \arctan\dfrac{b}{a}$.

If we use the four results (2.13) to (2.16), then r is always positive and the auxiliary angle α is always an acute angle.

Examples

1. $4\sin\theta + 3\cos\theta = r\sin(\theta+\alpha)$

$$r^2 = 4^2 + 3^2 = 25 \qquad \therefore\ r = 5$$

$$\alpha = \arctan\frac{3}{4} = \arctan 0.75 = 36°52'$$

$$\therefore\ \underline{4\sin\theta + 3\cos\theta = 5\sin(\theta + 36°52')}$$

2. $5\cos\theta - 2\sin\theta = r\cos(\theta+\alpha)$

$$r^2 = 5^2 + 2^2 = 25 + 4 = 29 \qquad \therefore\ r = 5.385$$

$$\tan\alpha = \frac{2}{5} = 0.4 \qquad \therefore\ \alpha = 21°48'$$

$$\therefore\ \underline{5\cos\theta - 2\sin\theta = 5.39\cos(\theta + 21°48')}$$

3. $7\sin\theta - 4\cos\theta = r\sin(\theta-\alpha)$

$$r^2 = 7^2 + 4^2 = 49 + 16 = 65 \qquad \therefore\ r = 8.062$$

$$\tan\alpha = \frac{4}{7} = 0.5714 \qquad \therefore\ \alpha = 29°45'$$

$$\therefore\ \underline{7\sin\theta - 4\cos\theta = 8.06\sin(\theta - 29°45')}$$

4. $3\cos 2\theta - \sin 2\theta = r\cos(2\theta+\alpha)$

$$r^2 = 3^2 + 1^2 = 10 \qquad \therefore\ r = 3.162$$

$$\tan\alpha = \frac{1}{3} = 0.3333 \qquad \therefore\ \alpha = 18°26'$$

$$\therefore\ \underline{3\cos 2\theta - \sin 2\theta = 3.16\cos(2\theta + 18°26')}$$

Exercise 6

Express each of the following as a sine or cosine of a compound angle.

1. $3\sin\theta + 5\cos\theta$
2. $6\sin\theta - 4\cos\theta$
3. $2\cos\theta + 3\sin\theta$
4. $5\cos\theta - 2\sin\theta$
5. $4\sin 2\theta - 3\cos 2\theta$
6. $2.6\cos 5\theta - 1.4\sin 5\theta$
7. $3.4\sin\theta + 2.1\cos\theta$
8. $4.63\cos 3\theta + 1.75\sin 3\theta$

2.2 HYPERBOLIC FUNCTIONS

2.2.1 Definitions

(a) $\sinh x = \dfrac{e^x - e^{-x}}{2}$

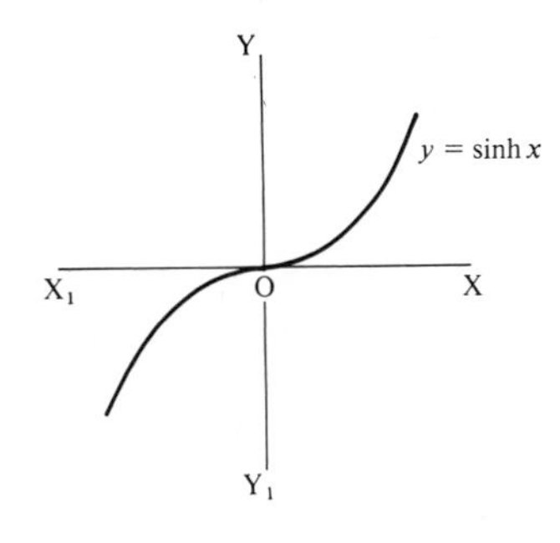

Note: (i) $\sinh 0 = 0$

(ii) curve symmetrical about the origin.

(b) $\cosh x = \dfrac{e^x + e^{-x}}{2}$

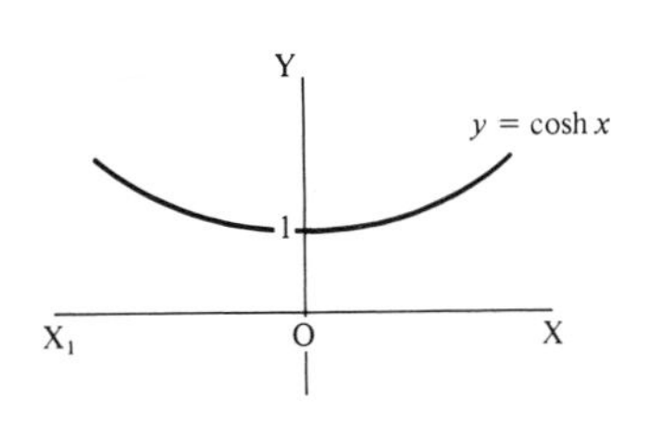

Note: (i) $\cosh 0 = 1$

(ii) curve symmetrical about the y-axis.

(c) $\tanh x = \dfrac{e^x - e^{-x}}{e^x + e^{-x}}$

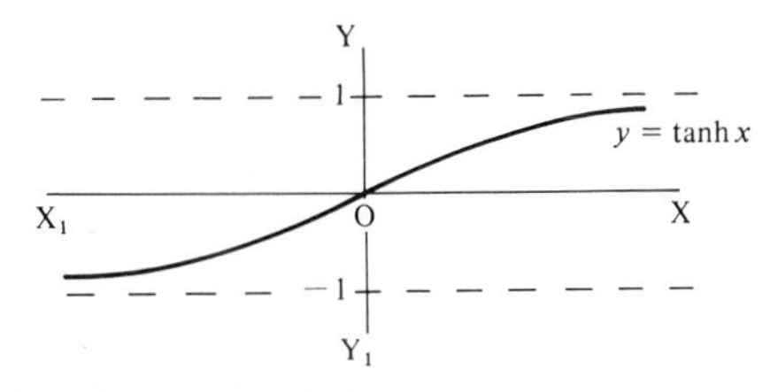

Note: (i) $\tanh 0 = 0$

(ii) curve symmetrical about the origin

(iii) as $x \to \pm\infty$, $\tanh x \to \pm 1$.

2.2.2 Hyperbolic identities

Hyperbolic identities bear a strong resemblance to those involving trigonometrical functions, except that a change of sign occurs whenever $\sin^2 x$ in the trigonometrical identity is being replaced by $\sinh^2 x$ in the corresponding hyperbolic identity, or is implied, or when a product of two sine terms is replaced by a product of two sinh terms.

	Trigonometrical identities	*Hyperbolic identities*
(a)	$\tan x = \dfrac{\sin x}{\cos x}$	$\tanh x = \dfrac{\sinh x}{\cosh x}$
	$\cot x = 1/\tan x$	$\coth x = 1/\tanh x$
	$\sec x = 1/\cos x$	$\operatorname{sech} x = 1/\cosh x$
	$\operatorname{cosec} x = 1/\sin x$	$\operatorname{cosech} x = 1/\sinh x$

(b) $\cos^2 x + \sin^2 x = 1$ $\qquad$ $\cosh^2 x \ominus \sinh^2 x = 1$

$\sec^2 x = 1 + \tan^2 x$ $\qquad$ $\text{sech}^2 x = 1 \ominus \tanh^2 x$

$\text{cosec}^2 x = 1 + \cot^2 x$ $\qquad$ $\text{cosech}^2 x = \coth^2 x \ominus 1$

(c) $\sin 2x = 2 \sin x \cos x$ $\qquad$ $\sinh 2x = 2 \sinh x \cosh x$

$$\cos 2x = \cos^2 x - \sin^2 x = 1 - 2\sin^2 x = 2\cos^2 x - 1$$

$$\cosh 2x = \cosh^2 x \oplus \sinh^2 x = 1 \oplus 2\sinh^2 x = 2\cosh^2 x - 1$$

$$\tan 2x = \frac{2 \tan x}{1 - \tan^2 x} \qquad \tanh 2x = \frac{2 \tanh x}{1 \oplus \tanh^2 x}$$

(d)

$$\sin(x+y) = \sin x \cos y + \cos x \sin y$$

$$\sin(x-y) = \sin x \cos y - \cos x \sin y$$

$$\cos(x+y) = \cos x \cos y - \sin x \sin y$$

$$\cos(x-y) = \cos x \cos y + \sin x \sin y$$

$$\sinh(x+y) = \sinh x \cosh y + \cosh x \sinh y$$

$$\sinh(x-y) = \sinh x \cosh y - \cosh x \sinh y$$

$$\cosh(x+y) = \cosh x \cosh y \oplus \sinh x \sinh y$$

$$\cosh(x-y) = \cosh x \cosh y \ominus \sinh x \sinh y$$

Note: The sign changes referred to at the beginning of this section have been ringed in the hyperbolic identities.

These identities will be useful later in the chapter dealing with the application of Fourier series to the solution of certain partial differential equations.

2.3 INTEGRATION BY PARTS

In the work that follows, it will often be necessary to integrate products of functions, e.g. $\int F_1(x)\, F_2(x)\, dx$ and this will be achieved by the method of 'integration by parts'.

$$\int u \frac{dv}{dx}\, dx = uv - \int v \frac{du}{dx}\, dx$$

This, for convenience, is sometimes written in the form

$$\int u\, dv = uv - \int v\, du$$

The product to be integrated consists of two functions

the first is denoted by 'u'

the second is regarded as being the differential coefficient of a further function 'v'.

To complete the right hand side,

(a) we know 'u'

(b) we find 'v' by integrating the function denoted as $\dfrac{dv}{dx}$

(c) we find $\dfrac{du}{dx}$ by differentiating the function denoted by u.

Example 1

To determine $\int x^2 e^{5x}\,dx$.

Let $\quad u = x^2 \qquad \therefore \dfrac{du}{dx} = 2x$

Let $\quad \dfrac{dv}{dx} = e^{5x} \qquad \therefore \quad v = \int e^{5x}\,dx = \dfrac{e^{5x}}{5}$

$$\therefore \int x^2 e^{5x}\,dx = x^2\left(\frac{e^{5x}}{5}\right) - \int \frac{e^{5x}}{5} 2x\,dx$$

$$= \frac{x^2e^{5x}}{5} - \frac{2}{5}\int x\,e^{5x}\,dx$$

Now repeat the process for the right hand integral.

Let $\quad u = x \qquad \therefore \dfrac{du}{dx} = 1$

Let $\quad \dfrac{dv}{dx} = e^{5x} \qquad \therefore \quad v = \int e^{5x}\,dx = \dfrac{e^{5x}}{5}$

$$\therefore \int x^2 e^{5x}\,dx = \frac{x^2e^{5x}}{5} - \frac{2}{5}\left\{\frac{xe^{5x}}{5} - \int \frac{e^{5x}}{5} 1\ dx\right\}$$

$$= \frac{x^2e^{5x}}{5} - \frac{2xe^{5x}}{25} + \frac{2e^{5x}}{125} + C$$

$$= \frac{e^{5x}}{5}\left\{x^2 - \frac{2x}{25} + \frac{2}{25}\right\} + C$$

Priority for 'u'

Where both factors are integrable, it is important to choose the correct one for 'u'. In the example above, we took x^2 to be u and this reduced to x in the second integral and to 1 in the final integral. Had we taken e^{5x} as u, then x^2 would have been $\frac{dv}{dx}$ and, on integration, this would have become $\frac{x^3}{3}$, the power of x getting successively greater, without eventually reducing to the factor 1.

A useful guide is a priority list for 'u'

(a) $\log nx$ (b) x^n (c) e^{kx}

Some special cases require special treatment, but this list applies to a large number of cases in practice.

Example 2

To determine $\int x^2 \ln x \, dx$.

Note that a log function is higher up the list than x^2.

Therefore, let

$$u = \ln x \qquad \therefore \frac{du}{dx} = \frac{1}{x}$$

$$\frac{dv}{dx} = x^2 \qquad \therefore v = \int x^2 \, dx = \frac{x^3}{3}$$

$$\therefore \int x^2 \ln x \, dx = \ln x \frac{x^3}{3} - \frac{1}{3}\int x^3 \frac{1}{x} \, dx$$

$$= \frac{x^3}{3} \ln x - \frac{1}{3}\int x^2 \, dx$$

$$= \frac{x^3}{3} \ln x - \frac{x^3}{9} + C$$

$$= \frac{x^3}{3}\left\{\ln x - \frac{1}{3}\right\} + C$$

Example 3

To determine $\int x \sin 2x \, dx$.

Let $u = x \qquad \therefore \frac{du}{dx} = 1$

$$\frac{dv}{dx} = \sin 2x \qquad \therefore \quad v = -\frac{\cos 2x}{2}$$

$$\therefore \int x \sin 2x \, dx = x\left(-\frac{\cos 2x}{2}\right) + \frac{1}{2}\int \cos 2x \times 1 \, dx$$

$$= -\frac{x \cos 2x}{2} + \frac{\sin 2x}{4} + C$$

Example 4

To determine $\int e^{5x} \cos 3x \, dx$.

Let $\quad u = e^{5x} \qquad \therefore \frac{du}{dx} = 5e^{5x}$

$$\frac{dv}{dx} = \cos 3x \qquad \therefore \quad v = \int \cos 3x \, dx = \frac{\sin 3x}{3}$$

$$\therefore \int e^{5x} \cos 3x \, dx = e^{5x}\left(\frac{\sin 3x}{3}\right) - \frac{5}{3}\int (\sin 3x) e^{5x} \, dx$$

$$= \frac{e^{5x} \sin 3x}{3} - \frac{5}{3}\int e^{5x} \sin 3x \, dx$$

For the right hand integral.

Let $\quad u = e^{5x} \qquad \therefore \frac{du}{dx} = 5e^{5x}$

$$\frac{dv}{dx} = \sin 3x \qquad \therefore \quad v = \frac{-\cos 3x}{3}$$

$$\therefore \int e^{5x} \cos 3x \, dx = \frac{e^{5x} \sin 3x}{3} - \frac{5}{3}\left\{e^{5x}\left(\frac{-\cos 3x}{3}\right) + \frac{5}{3}\int (\cos 3x) e^{5x} \, dx\right\}$$

$$= \frac{e^{5x} \sin 3x}{3} + \frac{5}{9} e^{5x} \cos 3x - \frac{25}{9}\int e^{5x} \cos 3x \, dx$$

i.e. $$I = \frac{e^{5x}}{3}\left\{\sin 3x + \frac{5 \cos 3x}{3}\right\} - \frac{25}{9} I$$

$$\therefore \frac{34}{9} I = \frac{e^{5x}}{3}\left\{\sin 3x + \frac{5 \cos 3x}{3}\right\}$$

$$\therefore I = \frac{3e^{5x}}{34}\left\{\sin 3x + \frac{5 \cos 3x}{3}\right\} + C$$

Example 5

To determine $\int \sin 3x \cos 5x \, dx$.

$$
\begin{aligned}
2 \sin A \cos B &= \sin (A+B) + \sin (A-B) \\
\therefore \quad \sin 3x \cos 5x &= \frac{1}{2}\{\sin (3x+5x) + \sin (3x-5x)\} \\
&= \frac{1}{2}\{\sin 8x - \sin 2x\}
\end{aligned}
$$

since $\sin(-2x) = -\sin 2x$

$$
\begin{aligned}
\therefore \quad \int \sin 3x \cos 5x \, dx &= \frac{1}{2}\int \{\sin 8x - \sin 2x\} \, dx \\
&= \frac{1}{2}\left\{-\frac{\cos 8x}{8} - \frac{\cos 2x}{2}\right\} + C \\
&= -\frac{1}{16}\{\cos 8x + 4\cos 2x\} + C
\end{aligned}
$$

Exercise 7

Determine the following.

1. $\int x^2 e^{3x} \, dx$
2. $\int x^3 \ln x \, dx$
3. $\int e^{2x} \sin x \, dx$
4. $\int x^2 \cos 2x \, dx$
5. $\int x e^{-5x} \, dx$
6. $\int t \sin \omega t \, dt$
7. $\int_{-\pi}^{\pi} (x+2) \sin 3x \, dx$
8. $\int_{0}^{\pi} (\theta^2 - 1) \cos 4\theta \, d\theta$
9. $\int_{0}^{\pi/2} \cos 4x \cos 7x \, dx$
10. $\int_{-\pi/6}^{\pi/6} \sin 6x \sin 2x \, dx$

Chapter 3

FOURIER SERIES FOR FUNCTIONS OF PERIOD 2π

3.1 PERIODIC FUNCTIONS

The essential characteristic of a periodic function is the fact that function values repeat regularly at a constant interval of the independent variable, i.e. the period of the function.

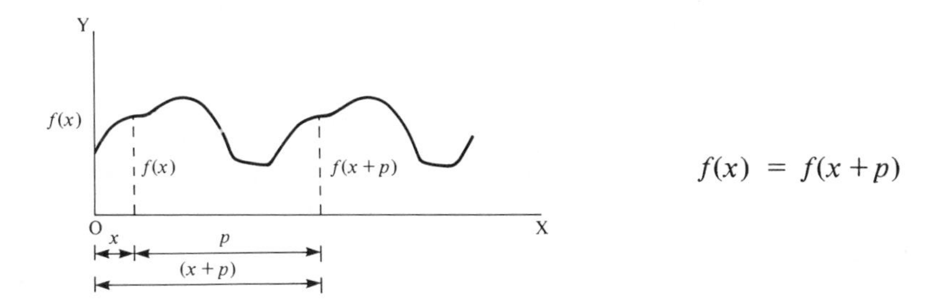

$$f(x) = f(x+p)$$

Since the pattern of the waveform repeats indefinitely, then it follows that $f(x) = f(x+np)$ where n is a positive integer, for all values of x.

Periodic functions of various kinds occur frequently in engineering and technological situations as mechanical vibrations, reciprocating action in linkages, electrical and electronic oscillations, and in problems relating to acoustics, etc. The study and analysis of periodic functions have been greatly facilitated by the work of Jean Baptiste Joseph Fourier (1768–1830) who showed that a function of a variable could be expanded as a series of sines of multiples of the variable. His theorem was later substantiated by Peter Gustav Lejeune Dirichlet (1805–59), provided that certain conditions were fulfilled.

3.2 FOURIER SERIES

3.2.1 General form of Fourier series

The basis of a Fourier series is to represent a function by a trigonometrical series of the form

$$\begin{aligned} f(x) = {} & A_0 + a_1 \cos x + a_2 \cos 2x + a_3 \cos 3x + \dots \\ & + b_1 \sin x + b_2 \sin 2x + b_3 \sin 3x + \dots \end{aligned}$$

For convenience in future calculations, we choose to re-write the constant term A_0 as $\frac{1}{2}a_0$, so in future we shall use the series in the form

$$f(x) = \tfrac{1}{2}a_0 + a_1 \cos x + a_2 \cos 2x + a_3 \cos 3x + \ldots + b_1 \sin x + b_2 \sin 2x + b_3 \sin 3x + \ldots$$

Since the sine and cosine terms have a period of 2π radians, it would appear that the Fourier series can represent only functions of period 2π. We shall deal later with periodic functions of other periods, but for the time being only functions having a period of 2π will be considered.

Therefore, *if $f(x)$ is a periodic function of period 2π*, the Fourier series to represent $f(x)$ can be expressed as

$$f(x) = \tfrac{1}{2}a_0 + a_1 \cos x + a_2 \cos 2x + a_3 \cos 3x + \ldots + b_1 \sin x + b_2 \sin 2x + b_3 \sin 3x + \ldots$$

where $a_0, a_1, a_2, a_3, \ldots, b_1, b_2, b_3, \ldots,$ are called the *Fourier coefficients* of $f(x)$.

The series can thus be written in the form

$$f(x) = \tfrac{1}{2}a_0 + \sum_{n=1}^{\infty} a_n \cos nx + \sum_{n=1}^{\infty} b_n \sin nx$$

i.e.

$$f(x) = \tfrac{1}{2}a_0 + \sum_{n=1}^{\infty} \{a_n \cos nx + b_n \sin nx\}$$

(n a positive integer)

3.2.2 Dirichlet conditions

The question of the validity of a Fourier series to represent a given function was investigated by Dirichlet who listed the conditions set out below.

If the Fourier series is to represent the function $f(x)$, then substituting $x = x_1$ will give an infinite series in x_1, the value of which should converge to the value of $f(x_1)$ as the number of terms considered is increased. For this to be achieved, the following conditions must be fulfilled throughout the periodic interval.

(a) *$f(x)$ must be defined and single-valued*, i.e. for each value of x there is one and only one value of $f(x)$.

For example, $y = x^2$ satisfies the condition,

but $y^2 = x$, i.e. $y = \pm\sqrt{x}$, does not.

(b) *$f(x)$ must be continuous or have a finite number of finite discontinuities within the periodic interval.*

For example

(i)

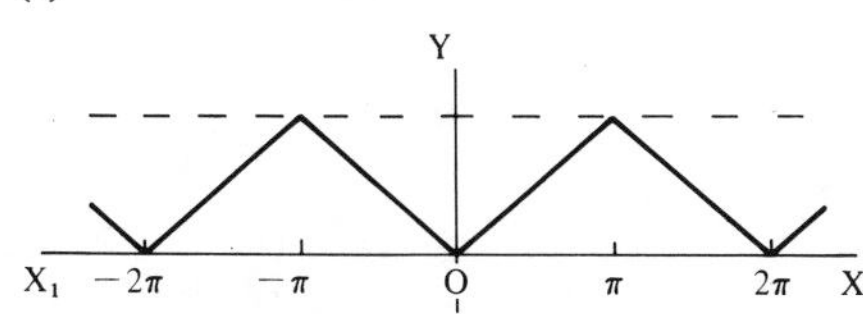

continuous (no breaks or jumps) within the cycle

(ii)

finite number of finite discontinuities (or jumps) within the cycle

∴ (i) and (ii) satisfy the condition.

(iii)

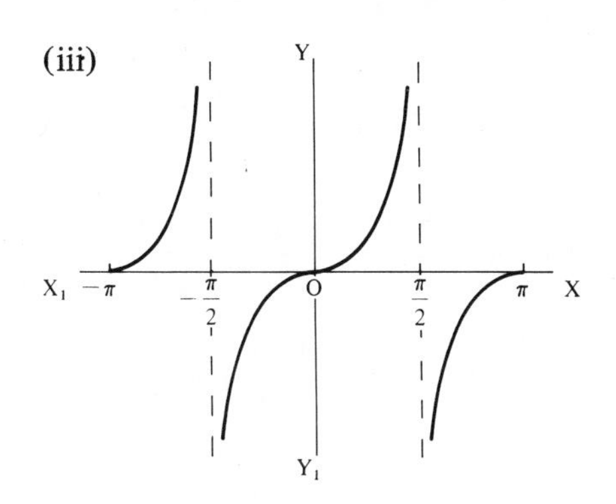

infinite jumps, i.e. non-finite discontinuities

∴ (iii) does not satisfy the condition.

(c) *$f(x)$ and $f'(x)$ are piecewise continuous in the periodic interval*, i.e. for sub-intervals of the periodic interval.

If these three conditions are satisfied, then the Fourier series converges to $f(x_1)$ if $x = x_1$ is a point of continuity.

In the majority of cases which occur in technological situations, these conditions are met and the series can be assumed faithfully to represent the particular function.

Exercise 8

In each of the following cases, state whether or not the function can be represented by a Fourier series. Each function is defined over the interval $-\pi < x < \pi$. and $f(x) = f(x + 2\pi)$.

1. $f(x) = x^2$
2. $f(x) = 3x + 4$
3. $f(x) = \dfrac{1}{x}$
4. $f(x) = \dfrac{1}{x+5}$

5. $f(x) = \tan x$
6. $f(x) = y$ where $x^2 + y^2 = 4$
7. $f(x) = \cos^2 x$
8. $f(x) = \sec x$

3.3 USEFUL INTEGRALS

We shall need to make frequent use of the results of the following integrals. In each case, n and m are integers other than zero.

(a) $$\int_{-\pi}^{\pi} \sin nx \, dx = \left[\frac{-\cos nx}{n}\right]_{-\pi}^{\pi} = \frac{1}{n}\{-\cos n\pi + \cos(-n\pi)\} \qquad \cos(-\theta) = \cos\theta$$
$$= \frac{1}{n}\{-\cos n\pi + \cos n\pi\} = \underline{0}$$

(b) $$\int_{-\pi}^{\pi} \cos nx \, dx = \left[\frac{\sin nx}{n}\right]_{-\pi}^{\pi} = \frac{1}{n}\{\sin n\pi - \sin(-n\pi)\} \qquad \sin(-\theta) = -\sin\theta$$
$$= \frac{1}{n}\{\sin n\pi + \sin n\pi\} = \underline{0}$$

(c) $$\int_{-\pi}^{\pi} \sin^2 nx \, dx = \frac{1}{2}\int_{-\pi}^{\pi} (1 - \cos 2nx) \, dx = \frac{1}{2}\left[x - \frac{\sin 2nx}{2n}\right]_{-\pi}^{\pi}$$
$$= \frac{1}{2}\{(\pi - 0) - (-\pi - 0)\} = \frac{1}{2}\{2\pi\} = \underline{\pi} \qquad (n \neq 0)$$

(d) $$\int_{-\pi}^{\pi} \cos^2 nx \, dx = \frac{1}{2}\int_{-\pi}^{\pi} (1 + \cos 2nx) \, dx = \frac{1}{2}\left[x + \frac{\sin 2nx}{2n}\right]_{-\pi}^{\pi}$$
$$= \frac{1}{2}\{(\pi + 0) - (-\pi + 0)\} = \frac{1}{2}\{2\pi\} = \underline{\pi} \qquad (n \neq 0)$$

(e) $$\int_{-\pi}^{\pi} \sin nx \cos mx \, dx = \frac{1}{2}\int_{-\pi}^{\pi} \{\sin(n+m)x + \sin(n-m)x\} dx$$
$$= \frac{1}{2}\{0 + 0\} = \underline{0} \qquad \text{from result (a) and } n \neq m$$

(f) $$\int_{-\pi}^{\pi} \cos nx \cos mx \, dx = \frac{1}{2}\int_{-\pi}^{\pi} \{\cos(n+m)x + \cos(n-m)x\} dx$$
$$= \frac{1}{2}\{0 + 0\} = \underline{0} \qquad \text{from result (b) and } n \neq m$$

If $n = m$ the integral becomes $\int_{-\pi}^{\pi} \cos^2 nx \, dx = \pi$ from result (d) above.

(g) $$\int_{-\pi}^{\pi} \sin nx \sin mx \, dx = \frac{1}{2}\int_{-\pi}^{\pi} \{\cos(n-m)x - \cos(n+m)x\} dx$$

$$= \frac{1}{2}\left[\frac{\sin(n-m)x}{n-m} - \frac{\sin(n+m)x}{n+m}\right]_{-\pi}^{\pi}$$

$$= \frac{1}{2}\{0-0\} = \underline{0} \qquad (n \neq m)$$

If $n = m$ the integral becomes $\int_{-\pi}^{\pi} \sin^2 nx \, dx = \pi$ from result (c) above.

Summary

Collecting the seven results together, we have

(a) $\int_{-\pi}^{\pi} \sin nx \, dx = 0$

(b) $\int_{-\pi}^{\pi} \cos nx \, dx = 0$

(c) $\int_{-\pi}^{\pi} \sin^2 nx \, dx = \pi \qquad (n \neq 0)$

(d) $\int_{-\pi}^{\pi} \cos^2 nx \, dx = \pi \qquad (n \neq 0)$

(e) $\int_{-\pi}^{\pi} \sin nx \cos mx \, dx = 0$

(f) $\int_{-\pi}^{\pi} \cos nx \cos mx \, dx = 0 \qquad (n \neq m)$

$= \pi \qquad (n = m)$

(g) $\int_{-\pi}^{\pi} \sin nx \sin mx \, dx = 0 \qquad (n \neq m)$

$= \pi \qquad (n = m)$

Provided integration is performed over an interval of 2π, the results are the same. The limits could, if required, be $-\pi$ to π; 0 to 2π; $-\frac{\pi}{2}$ to $\frac{3\pi}{2}$; $\frac{\pi}{2}$ to $\frac{5\pi}{2}$; etc.

3.4 FOURIER COEFFICIENTS

Earlier, we defined the Fourier series for a function $f(x)$ as

$$f(x) = \tfrac{1}{2}a_0 + a_1 \cos x + a_2 \cos 2x + a_3 \cos 3x + \ldots$$
$$+ b_1 \sin x + b_2 \sin 2x + b_3 \sin 3x + \ldots$$

i.e. $$f(x) = \tfrac{1}{2}a_0 + \sum_{n=1}^{\infty}\{a_n \cos nx + b_n \sin nx\} \qquad (n \text{ a positive integer})$$

To determine the series for any particular function, we have to find the numerical values of the Fourier coefficients.

(a) *To find* a_0, we integrate $f(x)$ with respect to x from $-\pi$ to π.

$$\int_{-\pi}^{\pi} f(x)\,dx = \frac{1}{2}\int_{-\pi}^{\pi} a_0\,dx + \sum_{n=1}^{\infty}\left\{\int_{-\pi}^{\pi} a_n \cos nx\,dx + \int_{-\pi}^{\pi} b_n \sin nx\,dx\right\}$$

$$= \frac{1}{2}\Big[a_0 x\Big]_{-\pi}^{\pi} + \sum_{n=1}^{\infty}\Big\{0+0\Big\} \qquad \text{from integrals (a) and (b) above}$$

$$= \frac{1}{2}\{a_0\pi - a_0(-\pi)\} = a_0\pi \qquad \therefore\ a_0 = \frac{1}{\pi}\int_{-\pi}^{\pi} f(x)\,dx$$

(b) *To find* a_n, we multiply $f(x)$ by $\cos mx$ and integrate from $-\pi$ to π.

$$\int_{-\pi}^{\pi} f(x)\cos mx\,dx = \frac{1}{2}\int_{-\pi}^{\pi} a_0 \cos mx\,dx + \sum_{n=1}^{\infty}\left\{\int_{-\pi}^{\pi} a_n \cos nx \cos mx\,dx + \int_{-\pi}^{\pi} b_n \sin nx \cos mx\,dx\right\}$$

$$= \frac{1}{2}a_0\times 0 + \sum_{n=1}^{\infty}\{a_n\times 0 + b_n\times 0\} \qquad \text{for } n \neq m$$

$$= \quad 0 \quad + \quad a_n\pi \quad + \quad 0 \qquad \text{for } n = m$$

$$= a_n\pi \qquad \therefore\ a_n = \frac{1}{\pi}\int_{-\pi}^{\pi} f(x)\cos nx\,dx$$

(c) *To find* b_n, we multiply $f(x)$ by $\sin mx$ and integrate from $-\pi$ to π.

$$\int_{-\pi}^{\pi} f(x)\sin mx\,dx = \frac{1}{2}\int_{-\pi}^{\pi} a_0 \sin mx\,dx + \sum_{n=1}^{\infty}\left\{\int_{-\pi}^{\pi} a_n \cos nx \sin mx\,dx + \int_{-\pi}^{\pi} b_n \sin nx \sin mx\,dx\right\}$$

$$= \frac{1}{2}a_0\times 0 + \sum_{n=1}^{\infty}\{a_n\times 0 + b_n\times 0\} \qquad \text{for } n \neq m$$

$$= 0 + a_n 0 + b_n \pi \qquad \text{for } n = m$$

$$= b_n \pi \qquad \therefore\ b_n = \frac{1}{\pi}\int_{-\pi}^{\pi} f(x) \sin nx\, dx$$

So we have

(a) $a_0 = \frac{1}{\pi}\int_{-\pi}^{\pi} f(x)\, dx$ i.e. $2 \times$ mean value of $f(x)$ over a period

(b) $a_n = \frac{1}{\pi}\int_{-\pi}^{\pi} f(x) \cos nx\, dx$ i.e. $2 \times$ mean value of $f(x) \cos nx$ over a period

(c) $b_n = \frac{1}{\pi}\int_{-\pi}^{\pi} f(x) \sin nx\, dx$ i.e. $2 \times$ mean value of $f(x) \sin nx$ over a period

for $n = 1, 2, 3, \ldots$

Example 1

Determine the Fourier series to represent the periodic function shown.

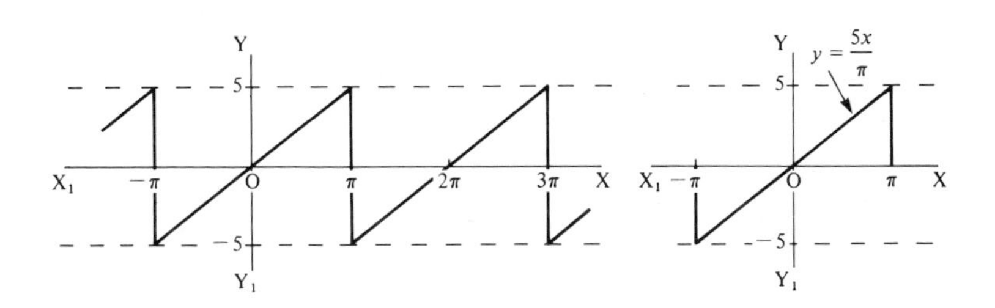

The function can be defined as $f(x) = \frac{5}{\pi}x, \ -\pi < x < \pi$.

(a) $a_0 = \frac{1}{\pi}\int_{-\pi}^{\pi} f(x)\, dx = \frac{1}{\pi}\int_{-\pi}^{\pi} \frac{5x}{\pi} dx = 0$ since $\int_{-\pi}^{\pi} f(x)\, dx$ evaluates the area between the line $y = \frac{5x}{\pi}$ and the x-axis. This consists of two equal parts, one positive, the other negative. Therefore the total area is zero.

$$\therefore\ a_0 = 0$$

(b) $a_n = \frac{1}{\pi}\int_{-\pi}^{\pi} f(x) \cos nx\, dx = \frac{1}{\pi}\int_{-\pi}^{\pi} \frac{5x}{\pi} \cos nx\, dx = \frac{5}{\pi^2}\int_{-\pi}^{\pi} x \cos nx\, dx.$

Integrating by parts, this gives

$$a_n = \frac{5}{\pi^2}\left\{\left[x\frac{\sin nx}{n}\right]_{-\pi}^{\pi} - \frac{1}{n}\int_{-\pi}^{\pi} \sin nx \, dx\right\}$$

$$= \frac{5}{\pi^2}\left\{\left[\frac{x \sin nx}{n}\right]_{-\pi}^{\pi} + \frac{1}{n}\times 0\right\} \qquad \text{(from the established integrals)}$$

$$= \frac{5}{\pi^2}\{0-0\} = 0 \qquad \therefore\ a_n = 0$$

(c) $b_n = \dfrac{1}{\pi}\displaystyle\int_{-\pi}^{\pi} f(x) \sin nx \, dx = \dfrac{1}{x}\displaystyle\int_{-\pi}^{\pi} \dfrac{5x}{\pi} \sin nx \, dx.$

Integrating by parts

$$b_n = \frac{5}{\pi^2}\left\{\left[x\left(\frac{-\cos nx}{n}\right)\right]_{-\pi}^{\pi} + \frac{1}{n}\int_{-\pi}^{\pi} \cos nx \, dx\right\}$$

$$= \frac{5}{\pi^2}\left\{\frac{-\pi \cos n\pi}{n} - \frac{\pi \cos n\pi}{n} + 0\right\} \qquad \text{(from the established integrals)}$$

$$= -\frac{10}{\pi}\frac{\cos n\pi}{n} \qquad \cos n\pi = 1 \quad (n \text{ even}) \quad = -1 \quad (n \text{ odd})$$

$$\therefore\ b_n = -\frac{10}{n\pi} \qquad (n \text{ even})$$

$$= \frac{10}{n\pi} \qquad (n \text{ odd})$$

We can combine these two expressions in the form

$$b_n = \frac{10}{n\pi}(-1)^{n+1}$$

In the Fourier series for $f(x)$, therefore,

$$a_0 = 0; \qquad a_n = 0; \qquad b_n = \frac{10}{n\pi}(-1)^{n+1}$$

In this particular case, the constant term and the coefficients of the cosine terms are all zero. The series therefore consists entirely of sine terms.

$$\therefore\ f(x) = \frac{10}{\pi}\left\{\sin x - \frac{1}{2}\sin 2x + \frac{1}{3}\sin 3x - \frac{1}{4}\sin 4x + \ldots\right\}$$

Example 2

Determine the Fourier series for the periodic function shown.

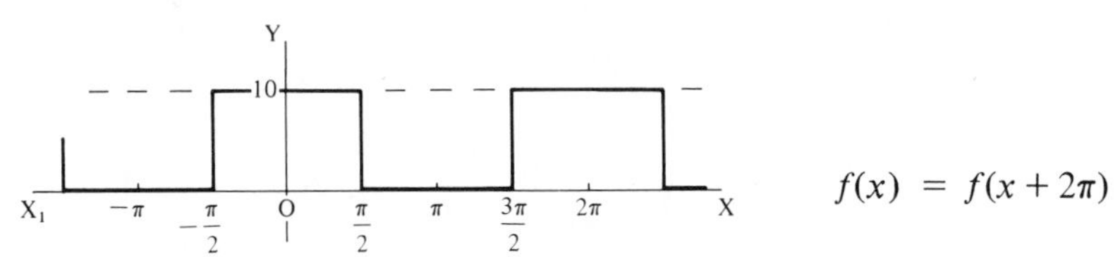

$$f(x) = f(x+2\pi)$$

Consider the values between $x=-\pi$ and $x=\pi$.

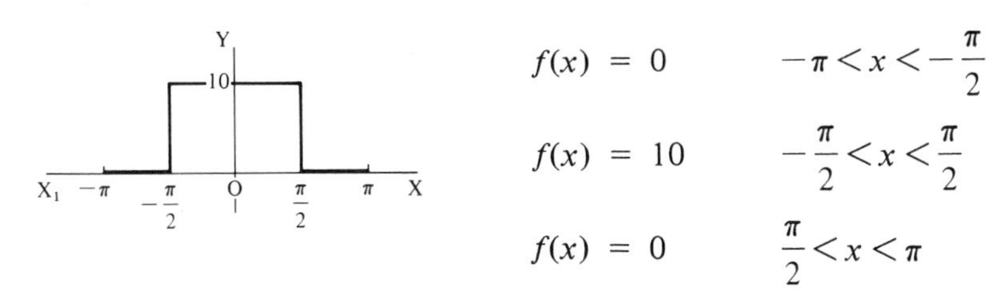

$$f(x) = 0 \qquad -\pi < x < -\frac{\pi}{2}$$

$$f(x) = 10 \qquad -\frac{\pi}{2} < x < \frac{\pi}{2}$$

$$f(x) = 0 \qquad \frac{\pi}{2} < x < \pi$$

Now
$$f(x) = \tfrac{1}{2}a_0 + \sum_{n=1}^{\infty}\{a_n \cos nx + b_n \sin nx\}$$

where
$$a_0 = \frac{1}{\pi}\int_{-\pi}^{\pi} f(x)\,dx; \qquad a_n = \frac{1}{\pi}\int_{-\pi}^{\pi} f(x)\cos nx\,dx$$

$$b_n = \frac{1}{\pi}\int_{-\pi}^{\pi} f(x)\sin nx\,dx$$

(a) *To find a_0*

$$\pi a_0 = \int_{-\pi}^{\pi} f(x)\,dx = \int_{-\pi}^{-\pi/2} 0\,dx + \int_{-\pi/2}^{\pi/2} 10\,dx + \int_{\pi/2}^{\pi} 0\,dx$$

$$= 10\Big[x\Big]_{-\pi/2}^{\pi/2} = 10\left\{\frac{\pi}{2}+\frac{\pi}{2}\right\} = 10\pi \qquad \therefore\ \underline{a_0 = 10}$$

$$\therefore\ \underline{\tfrac{1}{2}a_0 = 5}$$

(b) *To find a_n*

$$\pi a_n = \int_{-\pi/2}^{\pi} f(x)\ \cos nx\,dx$$

$$= \int_{-\pi}^{-\pi/2} 0\cos nx\,dx + \int_{-\pi/2}^{\pi/2} 10\cos nx\,dx + \int_{\pi/2}^{\pi} 0\cos nx\,dx$$

$$= 10\left[\frac{\sin nx}{n}\right]_{-\pi/2}^{\pi/2} = \frac{10}{n}\left\{\sin\frac{n\pi}{2} + \sin\frac{n\pi}{2}\right\}$$

$$= \frac{20}{n}\sin\frac{n\pi}{2}$$

$$\therefore \underline{a_n = \frac{20}{n\pi}\sin\frac{n\pi}{2}} = 0 \qquad \text{for } n \text{ even}$$

$$= \frac{20}{n\pi} \qquad \text{for } n = 1, 5, 9, \ldots$$

$$= -\frac{20}{n\pi} \qquad \text{for } n = 3, 7, 11, \ldots$$

(c) *To find b_n*

$$\pi b_n = \int_{-\pi}^{\pi} f(x)\sin nx\,\mathrm{d}x$$

$$= \int_{-\pi}^{-\pi/2} 0\sin nx\,\mathrm{d}x + \int_{-\pi/2}^{\pi/2} 10\sin nx\,\mathrm{d}x + \int_{\pi/2}^{\pi} 0\sin nx\,\mathrm{d}x$$

$$= 10\left[\frac{-\cos nx}{n}\right]_{-\pi/2}^{\pi/2} = -\frac{10}{n}\left\{\cos\frac{n\pi}{2} - \cos\frac{n\pi}{2}\right\} = 0$$

$$\therefore \underline{b_n = 0}$$

Therefore we have

$$a_0 = 10$$

$$a_n = 0 \qquad \text{for } n \text{ even}$$

$$= \frac{20}{n\pi} \qquad \text{for } n = 1, 5, 9, \ldots$$

$$= -\frac{20}{n\pi} \qquad \text{for } n = 3, 7, 11, \ldots$$

$$b_n = 0$$

In this case, there are no sine terms in the required series

$$\therefore f(x) = \tfrac{1}{2}a_0 + a_1\cos x + a_2\cos 2x + a_3\cos 3x + \ldots$$

$$+ b_1\sin x + b_2\sin 2x + b_3\sin 3x + \ldots$$

$$= 5 + \frac{20}{\pi}\cos x - \frac{20}{3\pi}\cos 3x + \frac{20}{5\pi}\cos 5x - \frac{20}{7\pi}\cos 7x + \ldots$$

$$+ \text{(no sine terms)}$$

$$\therefore \underline{f(x) = 5 + \frac{20}{\pi}\left\{\cos x - \frac{1}{3}\cos 3x + \frac{1}{5}\cos 5x - \frac{1}{7}\cos 7x + \ldots\right\}}$$

Example 3

Find the Fourier series for the function defined by

$$f(x) = 0 \qquad -\pi < x < 0$$
$$f(x) = x \qquad 0 < x < \pi$$
$$f(x) = f(x+2\pi)$$

First sketch the graph of the function.

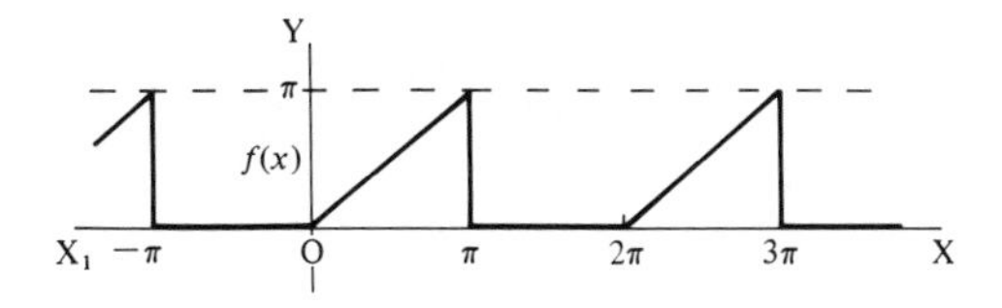

Considering the waveform between $x = -\pi$ and $x = \pi$,

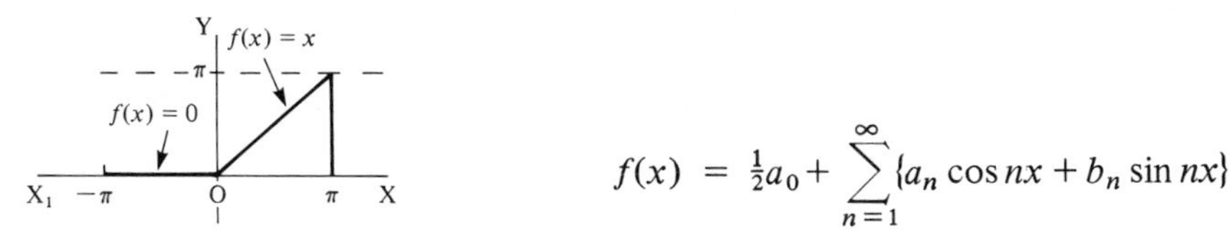

$$f(x) = \tfrac{1}{2}a_0 + \sum_{n=1}^{\infty}\{a_n \cos nx + b_n \sin nx\}$$

(a) *To find a_0*

$$a_0 = \frac{1}{\pi}\int_{-\pi}^{\pi} f(x)\,\mathrm{d}x$$

$$\therefore\ \pi a_0 = \int_{-\pi}^{0} 0\,\mathrm{d}x + \int_0^{\pi} x\,\mathrm{d}x = 0 + \left[\frac{x^2}{2}\right]_0^{\pi} = \frac{\pi^2}{2}$$

$$\therefore\ \underline{a_0 = \frac{\pi}{2}}$$

(b) *To find a_n*

$$a_n = \frac{1}{\pi}\int_{-\pi}^{\pi} f(x)\cos nx\,\mathrm{d}x$$

$$\therefore\ \pi a_n = \int_{-\pi}^{0} 0\cos nx\,\mathrm{d}x + \int_0^{\pi} x\cos nx\,\mathrm{d}x$$

$$= 0 + \left[x\left(\frac{\sin nx}{n}\right)\right]_0^{\pi} - \frac{1}{n}\int_0^{\pi}\sin nx\,\mathrm{d}x$$

$$= \left[\frac{x\sin nx}{n}\right]_0^{\pi} - \frac{1}{n}\left[\frac{-\cos nx}{n}\right]_0^{\pi}$$

$$= \frac{1}{n}(\pi \sin n\pi - 0) + \frac{1}{n^2}(\cos n\pi - \cos 0)$$

$$= \frac{\pi}{n} \sin n\pi + \frac{1}{n^2}(\cos n\pi - 1)$$

$$\sin n\pi = 0 \quad \text{for } n = 1, 2, 3, \ldots$$

$$\cos n\pi = -1 \quad \text{for } n = 1, 3, 5, \ldots$$

$$= 1 \quad \text{for } n = 2, 4, 6, \ldots$$

$$\therefore a_n = \frac{1}{n^2\pi}\{(-1)^n - 1\} = 0 \text{ for } n \text{ even}$$

$$= -\frac{2}{n^2\pi} \text{ for } n \text{ odd}$$

(c) *To find b_n*

$$b_n = \frac{1}{\pi}\int_{-\pi}^{\pi} f(x) \sin nx \, dx$$

$$\therefore \pi b_n = \int_{-\pi}^{0} 0 \sin nx \, dx + \int_0^{\pi} x \sin nx \, dx$$

$$= 0 + \left[x\left(\frac{-\cos nx}{n}\right)\right]_0^{\pi} + \frac{1}{n}\int_0^{\pi} \cos nx \, dx$$

$$= \left[-\frac{\pi}{n}\cos n\pi - 0\right] + \frac{1}{n^2}\left[\sin nx\right]_0^{\pi}$$

$$\sin n\pi = 0 \text{ for } n = 1, 2, 3, \ldots$$

$$= -\frac{\pi}{n}\cos n\pi + 0 \qquad \cos n\pi = 1 \text{ for } n \text{ even}$$

$$= -1 \text{ for } n \text{ odd}$$

$$\therefore b_n = -\frac{1}{n}(-1)^n$$

Therefore $a_0 = \frac{\pi}{2}$; $\quad a_n = 0$ for n even

$$= -\frac{2}{n^2\pi} \text{ for } n \text{ odd}$$

$$b_n = -\frac{1}{n}(-1)^n$$

$$f(x) = \frac{\pi}{4} - \frac{2}{\pi}\left\{\cos x + \frac{1}{3^2}\cos 3x + \frac{1}{5^2}\cos 5x + \ldots\right\}$$

$$+ \sin x - \frac{1}{2}\sin 2x + \frac{1}{3}\sin 3x - \frac{1}{4}\sin 4x + \ldots$$

Example 4

Obtain the Fourier series for the function defined by

$$f(x) = x(2\pi - x) \qquad 0 < x < 2\pi$$
$$f(x) = f(x + 2\pi)$$

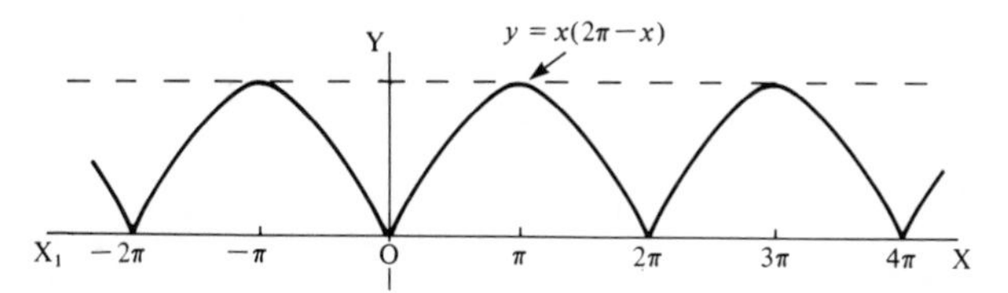

We established earlier that, so long as the interval of integration is 2π, the limits of the integration can be chosen as convenient. In this example, it would certainly be more convenient to take the limits as 0 and 2π rather than $-\pi$ and π as we have used previously. In the interval 0 to 2π, the function is continuous and the integration can be carried out in one stage.

(a) *To find* a_0

$$\pi a_0 = \int_0^{2\pi} f(x)\,dx = \int_0^{2\pi} x(2\pi - x)\,dx = \int_0^{2\pi} (2\pi x - x^2)\,dx$$

$$= \left[\pi x^2 - \frac{x^3}{3}\right]_0^{2\pi} = 4\pi^3 - \frac{8\pi^3}{3} = \frac{8\pi^3}{6}$$

$$\therefore\ a_0 = \frac{4\pi^2}{3}$$

(b) *To find* a_n

$$\pi a_n = \int_0^{2\pi} x(2\pi - x)\cos nx\,dx$$

$$= 2\pi\int_0^{2\pi} x\cos nx\,dx - \int_0^{2\pi} x^2\cos nx\,dx = I_1 - I_2$$

$$I_1 = 2\pi\left\{\left[x\left(\frac{\sin nx}{n}\right)\right]_0^{2\pi} - \frac{1}{n}\int_0^{2\pi}\sin nx\,dx\right\}$$

$$= 2\pi\left\{(0 - 0) \quad - \frac{1}{n}\times 0\right\} \qquad \therefore\ I_1 = 0$$

$$I_2 = \int_0^{2\pi} x^2\cos nx\,dx$$

$$= \left[x^2\left(\frac{\sin nx}{n}\right)\right]_0^{2\pi} - \frac{2}{n}\int_0^{2\pi} x\sin nx\,dx$$

$$= \left[\frac{x^2 \sin nx}{n}\right]_0^{2\pi} - \frac{2}{n}\left\{\left[x\left(\frac{-\cos nx}{n}\right)\right]_0^{2\pi} + \frac{1}{n}\int_0^{2\pi} \cos nx \, dx\right\}$$

$$= \quad 0 \quad + \quad \left[\frac{2x \cos nx}{n^2}\right]_0^{2\pi} \quad + \quad 0$$

$$= \frac{4\pi \cos 2n\pi}{n^2} = \frac{4\pi}{n^2} \qquad \therefore I_2 = \frac{4\pi}{n^2}$$

$$\therefore a_n = 0 - \frac{4}{n^2} \qquad \therefore a_n = -\frac{4}{n^2}$$

(c) *To find b_n*

$$\pi b_n = \int_0^{2\pi} x(2\pi - x) \sin nx \, dx$$

$$= \int_0^{2\pi} (2\pi x - x^2) \sin nx \, dx$$

$$= 2\pi \int_0^{2\pi} x \sin nx \, dx - \int_0^{2\pi} x^2 \sin nx \, dx = I_1 - I_2$$

$$I_1 = 2\pi \int_0^{2\pi} x \sin nx \, dx = 2\pi\left\{\left[x\left(\frac{-\cos nx}{n}\right)\right]_0^{2\pi} + \frac{1}{n}\int_0^{2\pi} \cos nx \, dx\right\}$$

$$= \frac{2\pi}{n}\left[-x \cos nx\right]_0^{2\pi} + 0$$

$$= \frac{2\pi}{n}[-2\pi \cos n2\pi] = -\frac{4\pi^2}{n} \qquad \therefore I_1 = -\frac{4\pi^2}{n}$$

$$I_2 = \int_0^{2\pi} x^2 \sin nx \, dx = \left[x^2\left(\frac{-\cos nx}{n}\right)\right]_0^{2\pi} + \frac{2}{n}\int_0^{2\pi} x \cos nx \, dx$$

$$= \left[\frac{-x^2 \cos nx}{n}\right]_0^{2\pi} + \frac{2}{n}\left\{\left[x\left(\frac{\sin nx}{n}\right)\right]_0^{2\pi} - \frac{1}{n}\int_0^{2\pi} \sin nx \, dx\right\}$$

$$= \left[\frac{-x^2 \cos nx}{n}\right]_0^{2\pi} + \quad 0 \quad - \quad 0$$

$$= -\frac{4\pi^2}{n} \qquad \therefore I_2 = -\frac{4\pi^2}{n}$$

$$\therefore \pi b_n = -\frac{4\pi^2}{n} + \frac{4\pi^2}{n} = 0 \qquad \therefore b_n = 0$$

$$\therefore\ a_0 = \frac{4\pi^2}{3}; \qquad a_n = -\frac{4}{n^2}; \qquad b_n = 0$$

$$\therefore\ f(x) = \frac{2\pi^2}{3} - 4\left\{\cos x + \frac{\cos 2x}{2^2} + \frac{\cos 3x}{3^2} + \frac{\cos 4x}{4^2} + \ldots\right\}$$

From the examples we have worked through, we see that in certain cases the Fourier series obtained contains only cosine terms or only sine terms, though, in general, both sine and cosine terms are present. We shall consider this point further later on.

Meanwhile, using the established results

$$f(x) = \tfrac{1}{2}a_0 + \sum_{n=1}^{\infty}\{a_n \cos nx + b_n \sin nx\}$$

where $$a_0 = \frac{1}{\pi}\int_{-\pi}^{\pi} f(x)\,\mathrm{d}x; \qquad a_n = \frac{1}{\pi}\int_{-\pi}^{\pi} f(x)\cos nx\,\mathrm{d}x$$

$$b_n = \frac{1}{\pi}\int_{-\pi}^{\pi} f(x)\ \sin nx\,\mathrm{d}x$$

and integrating by parts where appropriate, we can now determine the Fourier series for a number of different periodic functions.

Exercise 9

Establish the Fourier series for the functions defined below.

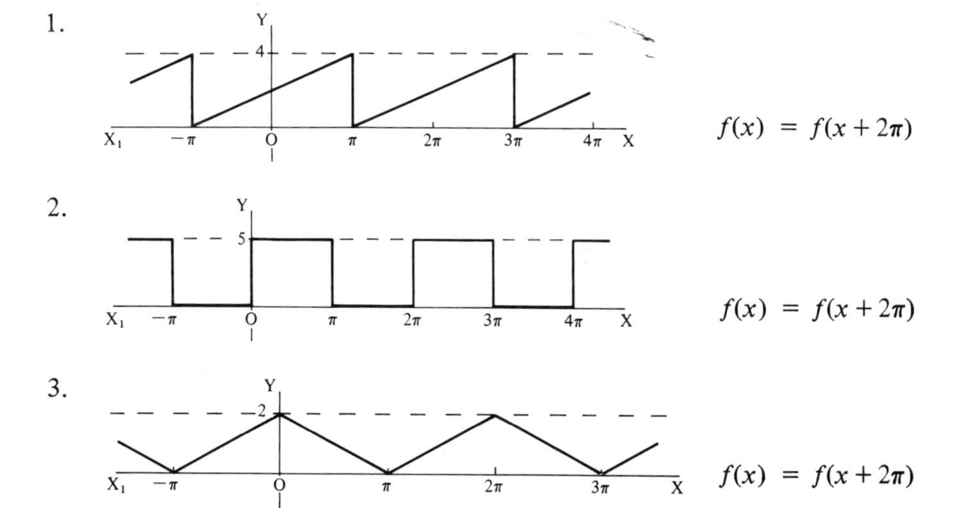

4.

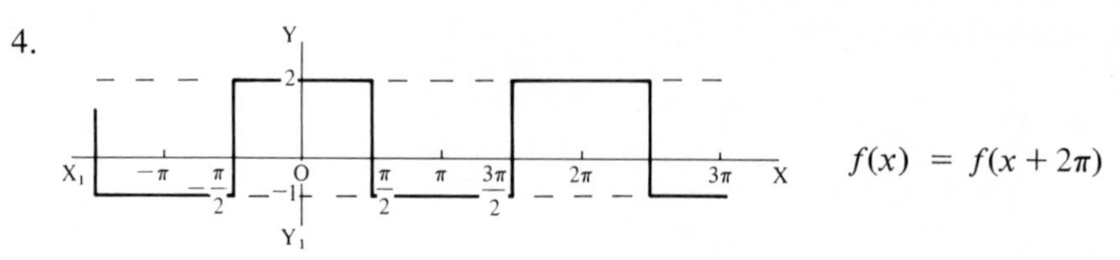

$f(x) = f(x + 2\pi)$

5. $f(x) = x + \pi \qquad -\pi < x < 0$

$f(x) = \pi \qquad 0 < x < \frac{\pi}{2}$

$f(x) = 0 \qquad \frac{\pi}{2} < x < \pi$

$f(x) = f(x + 2\pi)$

6. $f(x) = 6 \qquad -\pi < x < 0$

$f(x) = 2 \qquad 0 < x < \pi$

$f(x) = f(x + 2\pi)$

7.

$f(x) = f(x + 2\pi)$

8. $f(x) = -\pi \qquad -\pi < x < -\frac{\pi}{2}$

$f(x) = 2x \qquad -\frac{\pi}{2} < x < \frac{\pi}{2}$

$f(x) = \pi \qquad \frac{\pi}{2} < x < \pi$

$f(x) = f(x + 2\pi)$

9. $f(x) = -\frac{5}{\pi}x \qquad -\pi < x < 0$

$f(x) = \frac{5}{\pi}x \qquad 0 < x < \pi$

$f(x) = f(x + 2\pi)$

10. $f(x) = \frac{4x}{\pi} + 2 \qquad -\frac{\pi}{2} < x < \frac{\pi}{2}$

$f(x) = 4 \qquad \frac{\pi}{2} < x < \pi$

$f(x) = 0 \qquad \pi < x < \frac{3\pi}{2}$

$f(x) = f(x + 2\pi)$

3.5 CONVERGENCE OF A FOURIER SERIES

Convergence of the series for $f(x)$ to the value of $f(x_1)$ when $x = x_1$ necessarily means that the amplitudes of the higher order harmonics must eventually decrease in value as n increases, even though the amplitudes of the early harmonics may even increase,

i.e. as $n \to \infty$, $a_n \to 0$ and $b_n \to 0$

In practice, only the first few harmonics have any major effect on the function values and normally only the first six or so harmonics are needed.

3.6 FOURIER SERIES AT A DISCONTINUITY

If the function $f(x)$ has a finite discontinuity (i.e. a finite jump) at $x = x_1$, where x_1 lies in the periodic interval $-\pi < x_1 < \pi$, then at $x = x_1$ the function appears to have two values, y_1 and y_2. In fact, the function is not defined at $x = x_1$.

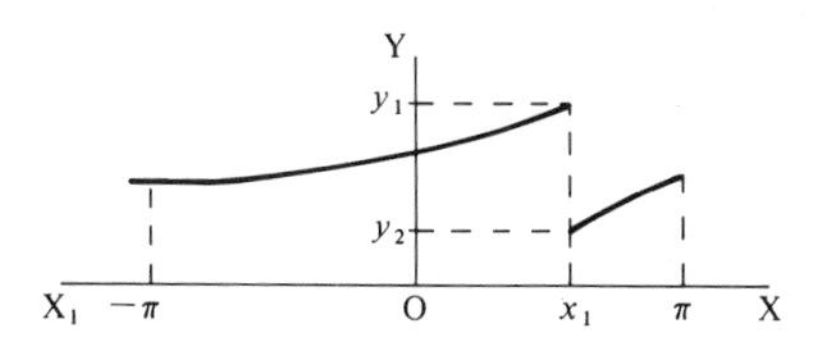

We can indicate this by writing

$$f(x_1 - 0) = y_1$$

and

$$f(x_1 + 0) = y_2$$

representing the facts that

(a) the limiting value of $f(x)$ as we approach x_1 from below is y_1

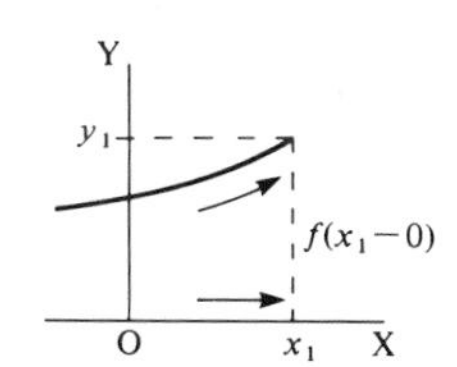

(b) the limiting value of $f(x)$ as we approach x_1 from above is y_2.

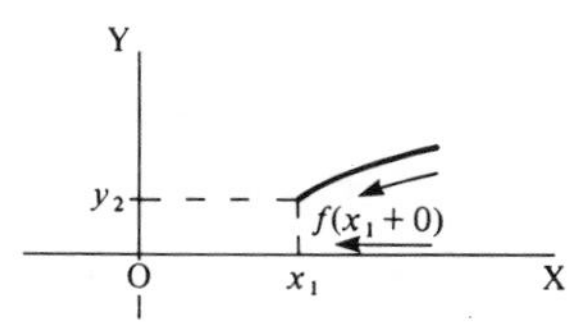

If we substitute $x = x_1$ in the Fourier series for $f(x)$, then it can be shown that the series converges to the value

$$\tfrac{1}{2}\{f(x_1 - 0) + f(x_1 + 0)\}$$

i.e. $\frac{1}{2}\{y_1 + y_2\}$, the average value of y_1 and y_2.

We can illustrate this by a simple example.

Example

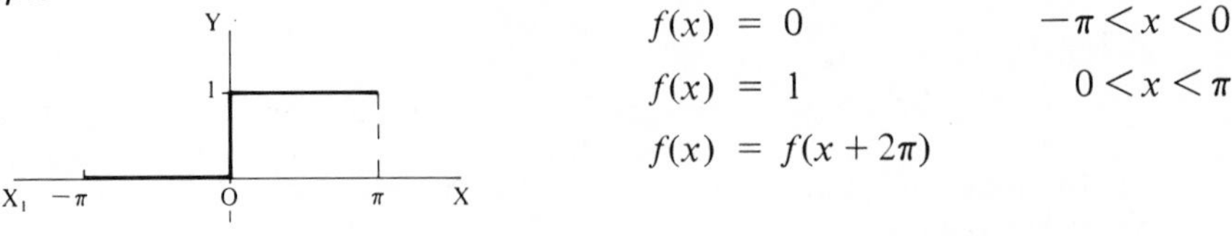

$$f(x) = 0 \qquad -\pi < x < 0$$
$$f(x) = 1 \qquad 0 < x < \pi$$
$$f(x) = f(x + 2\pi)$$

$$f(x) = \tfrac{1}{2}a_0 + \sum_{n=1}^{\infty} \{a_n \cos nx + b_n \sin nx\}$$

$$a_0 = \frac{1}{\pi}\left\{\int_{-\pi}^{0} 0 \,\mathrm{d}x + \int_{0}^{\pi} 1 \,\mathrm{d}x\right\} = \frac{1}{\pi}\left\{0 + \Big[x\Big]_0^{\pi}\right\} = \frac{1}{\pi}\pi = 1 \qquad \therefore \underline{a_0 = 1}$$

$$a_n = \frac{1}{\pi}\left\{\int_{-\pi}^{0} 0 \cos nx \,\mathrm{d}x + \int_{0}^{\pi} \cos nx \,\mathrm{d}x\right\} = \frac{1}{\pi}\left[\frac{\sin nx}{n}\right]_0^{\pi} = \frac{\sin n\pi}{n\pi}$$

$$\therefore \underline{a_n = 0}$$

$$b_n = \frac{1}{\pi}\left\{\int_{-\pi}^{0} 0 \sin nx \,\mathrm{d}x + \int_{0}^{\pi} \sin nx \,\mathrm{d}x\right\} = \frac{1}{\pi}\left[\frac{-\cos nx}{n}\right]_0^{\pi} = \frac{1 - \cos n\pi}{n\pi}$$

$$\therefore \underline{b_n = \frac{2}{n\pi}\,(n \text{ odd}) \qquad \text{and} \qquad 0\,(n \text{ even})}$$

$$f(x) = \frac{1}{2} + \frac{2}{\pi}\left\{\frac{1}{1}\sin x + \frac{1}{3}\sin 3x + \frac{1}{5}\sin 5x + \frac{1}{7}\sin 7x + \ldots\right\}$$

At $x = 0$, $\quad \sin nx = 0 \quad (n = 1, 2, 3, \ldots) \qquad \therefore \underline{f(x) = \frac{1}{2}}$

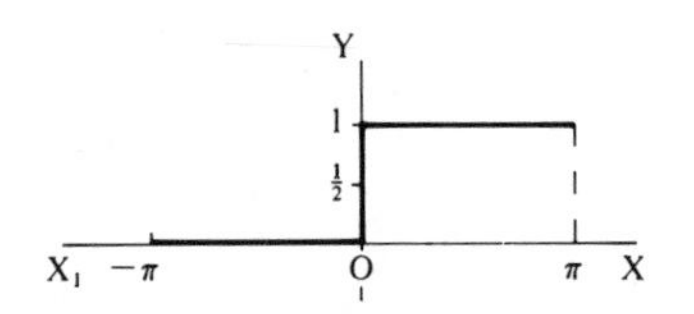

$\therefore$ At the discontinuity at $x = 0$ the series converges to the value $\frac{1}{2}$, i.e. the average value of $f(x_1 - 0)$ and $f(x_1 + 0)$ at $x_1 = 0$.

Therefore, the Fourier series for a function $f(x)$ converges to the value of

(a) $f(x_1)$ if the function is continuous at $x = x_1$

(b) $\dfrac{f(x_1 - 0) + f(x_1 + 0)}{2}$ if $x = x_1$ is a point of finite discontinuity.

3.7 ODD AND EVEN FUNCTIONS AND THEIR PRODUCTS

Our work with Fourier series can often be shortened by a knowledge of odd and even functions.

3.7.1 Even functions

A function $f(x)$ is said to be an *even* function if

$$f(-x) = f(x)$$

i.e. the function value for a particular negative value of x is identical to that for the corresponding positive value of x.

For example, $f(x) = x^2$ is an even function.

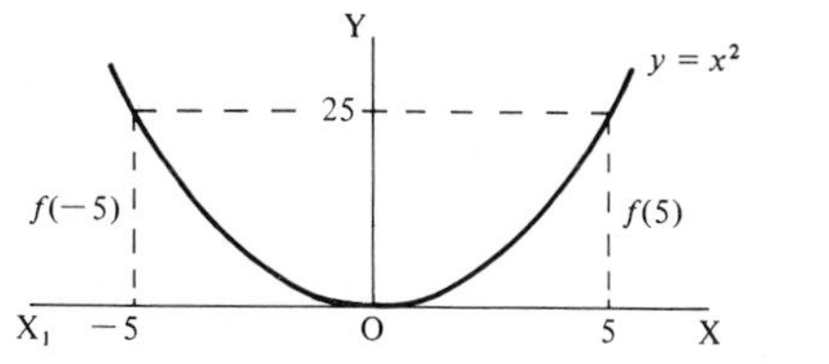

$f(-3) = 9$ and $f(3) = 9$,
i.e. $f(-3) = f(3)$

and

$f(-5) = 25$ and $f(5) = 25$,
i.e. $f(-5) = f(5)$

Similarly, $f(x) = \cos x$ is an even function, since for all values of x,

$$\cos(-x) = \cos x$$

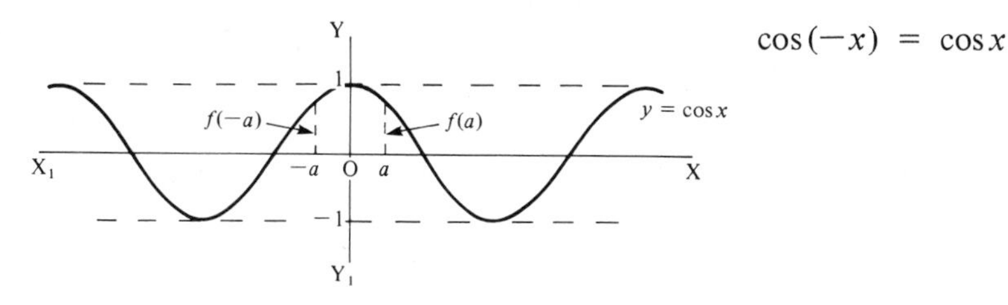

Note that an *even function* is necessarily *symmetrical about the y-axis.*

3.7.2 Odd functions

A function $f(x)$ is said to be an *odd* function if

$$f(-x) = -f(x)$$

i.e. the function value for a particular negative value of x is numerically the same as that for the corresponding positive value of x, but differs from it in sign.

For example, $f(x) = x^3$ is an odd function, since

$$\left.\begin{aligned} f(-2) &= (-2)^3 = -8 \\ f(2) &= (2)^3 \;\;\;= 8 \end{aligned}\right\} \quad \text{i.e. } f(-2) = -f(2)$$

and

$$\left.\begin{aligned} f(-5) &= (-5)^3 = -125 \\ f(5) &= (5)^3 \;\;\;= 125 \end{aligned}\right\} \quad \text{i.e. } f(-5) = -f(5)$$

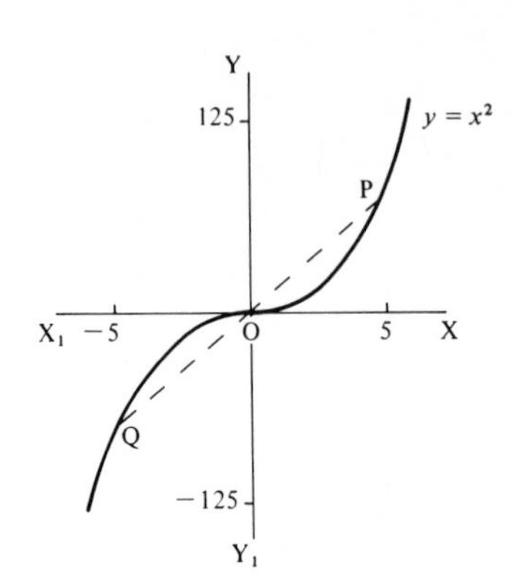

It follows that the graph of an *odd* function is *symmetrical about the origin.*

For every point P on the curve, there is another point Q on the curve on a straight line through the origin and equidistant from it.

Similarly, $f(x) = \sin x$ is an odd function since, for all values of x,

$$\sin(-x) = -\sin x$$

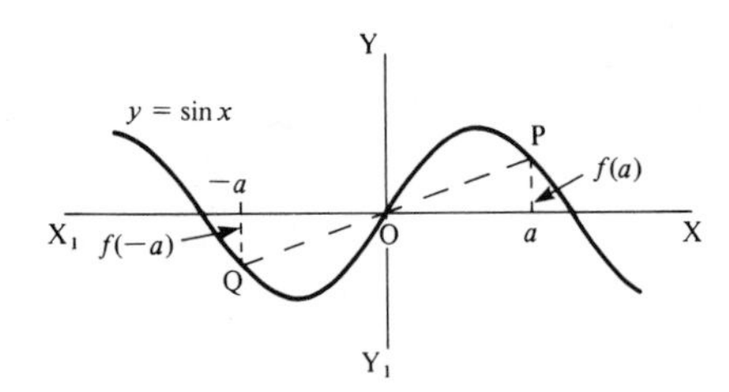

So, by the same means, $f(x) = \cos nx$ is an even function

and $f(x) = \sin nx$ is an odd function.

Summary

For an *even function*, $f(-x) = f(x)$

and the curve is symmetrical about the y-axis.

For an *odd function*, $f(-x) = -f(x)$

and the curve is symmetrical about the origin.

Exercise 10

For each waveform shown below, state whether the periodic function represented is an odd function, an even function, or neither.

1.

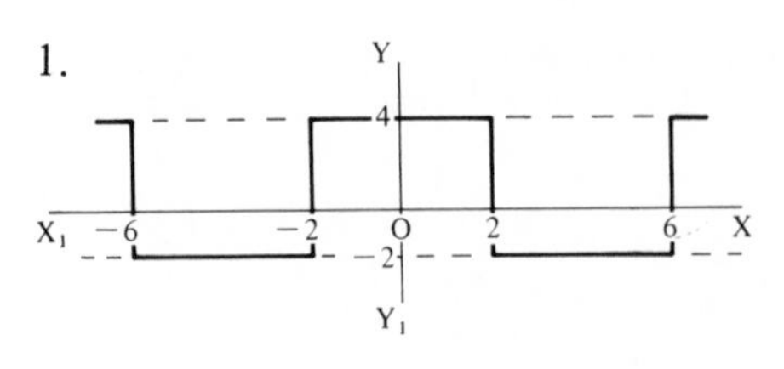

2.

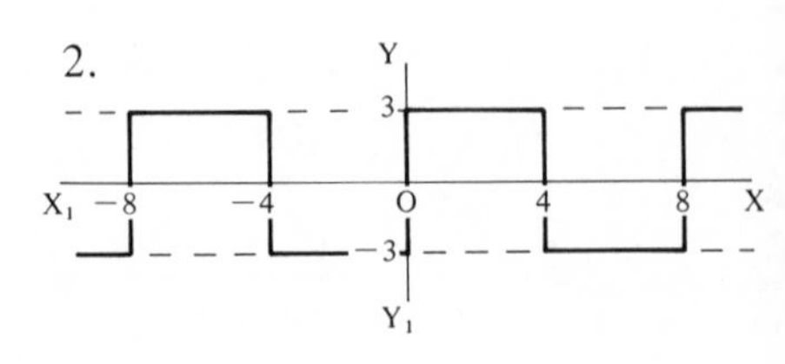

3.

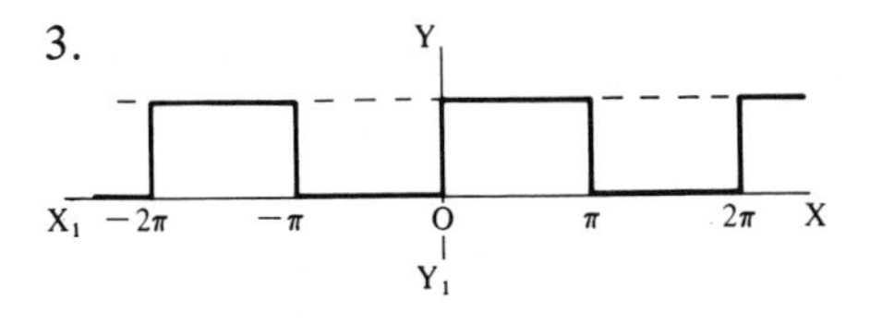

4.

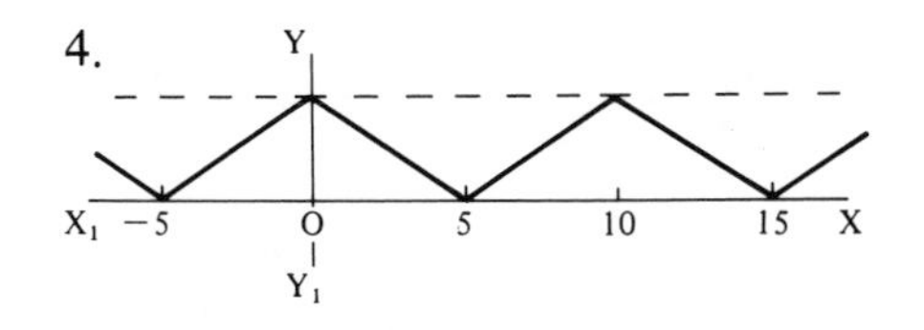

5.

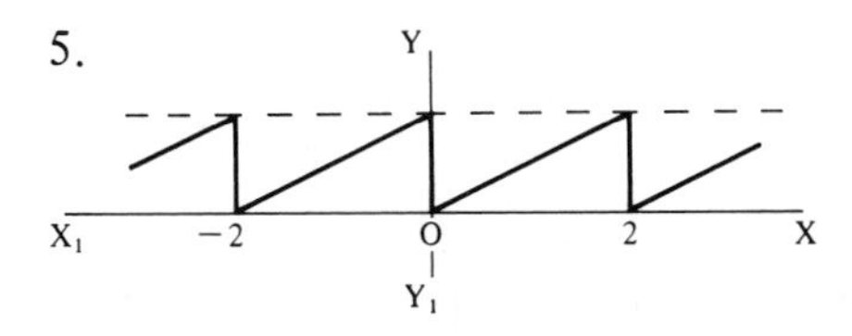

6.

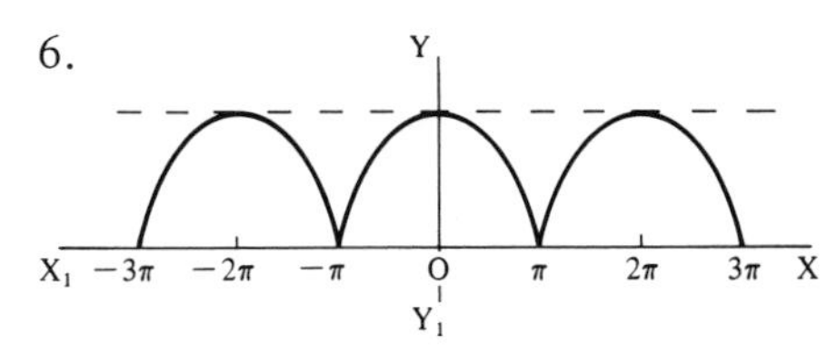

7.

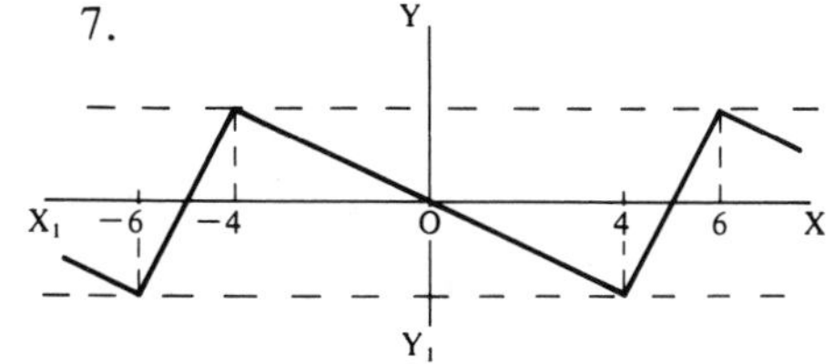

8.

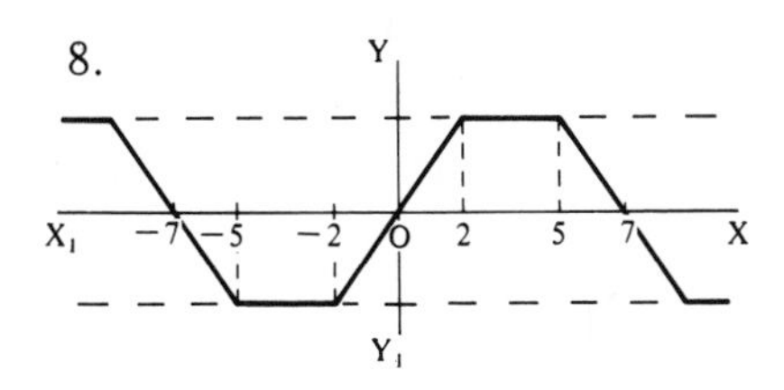

9.

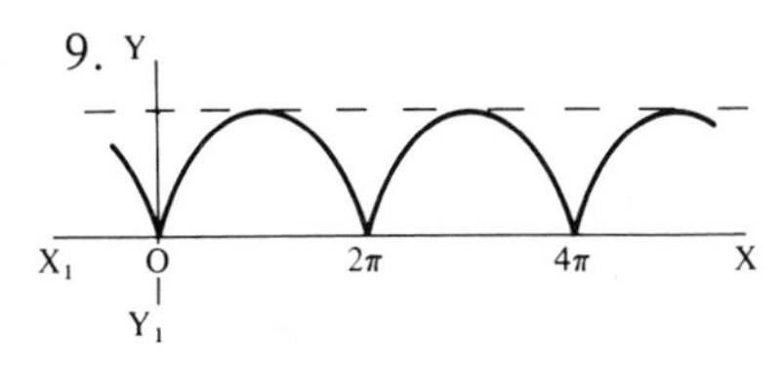

10.

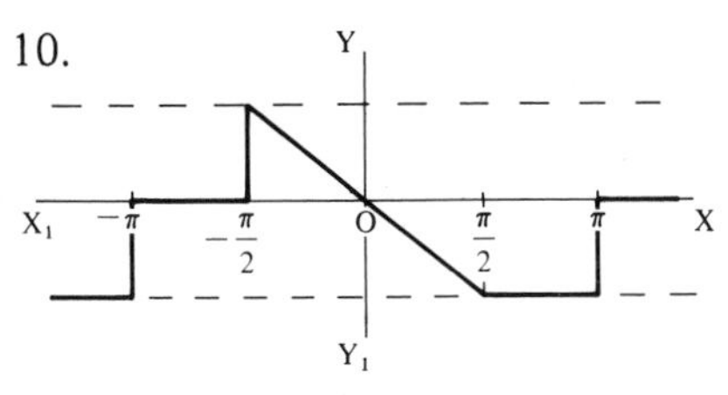

3.7.3 Products of odd and even functions

The odd or even nature of the product of two odd or even functions can be determined by a few simple rules which resemble very closely the elementary rules of signs.

$$\text{(even)} \times \text{(even)} = \text{(even)} \quad \text{corresponding to} \quad (+)\times(+) = (+)$$

$$\text{(odd)} \times \text{(odd)} = \text{(even)} \qquad (-)\times(-) = (+)$$

$$\text{(odd)} \times \text{(even)} = \text{(odd)} \qquad (-)\times(+) = (-)$$

These rules can easily be proved.

(a) Let $F(x) = f(x)\, g(x)$ where $f(x)$ and $g(x)$ are even functions

Then $F(-x) = f(-x)\, g(-x)$

$= f(x)\, g(x)$ since $f(-x) = f(x)$ and $g(-x) = g(x)$

$\therefore\ F(-x) = F(x)$ i.e. *F(x) is even*

(b) Let $F(x) = f(x)\,g(x)$ where $f(x)$ and $g(x)$ are odd functions

Then $F(-x) = f(-x)\,g(-x)$

$= \{-f(x)\}\,\{-g(x)\}$ since $f(-x) = -f(x)$ and $g(-x) = -g(x)$

$\therefore F(-x) = f(x)\,g(x) = F(x)$

$\therefore F(-x) = F(x)$ i.e. *F(x) is even*

(c) Let $F(x) = f(x)\,g(x)$ where $f(x)$ is even and $g(x)$ is odd

Then $F(-x) = f(-x)\,g(-x)$

$= f(x)\,\{-g(x)\}$ since $f(-x) = f(x)$ and $g(-x) = -g(x)$

$= -f(x)\,g(x) = -F(x)$

$\therefore F(-x) = -F(x)$ i.e. *F(x) is odd*

Therefore

if $f(x)$ and $g(x)$ are both *even*, then $f(x)\,g(x)$ is *even*

if $f(x)$ and $g(x)$ are both *odd*, then $f(x)\,g(x)$ is *even*

But, if either $f(x)$ or $g(x)$ is *even* and the other *odd*, then $f(x)\,g(x)$ is *odd*.

Note also

(a)

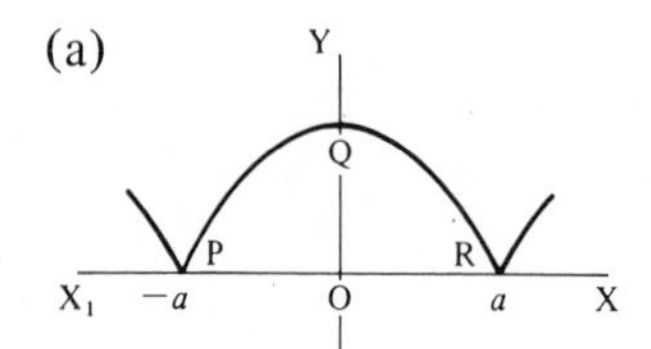

For an even function (symmetrical about the y-axis)

$$\int_{-a}^{a} f(x)\,dx = 2\int_{0}^{a} f(x)\,dx$$

since area PQO = area OQR.

(b)

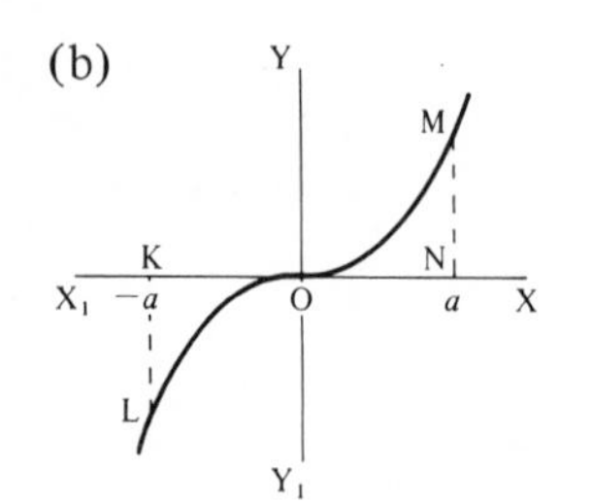

For an odd function (symmetrical about the origin)

$$\int_{-a}^{a} f(x)\,dx = 0$$

since area KLO = −area OMN.

These are useful facts to remember.

For example, consider $\int_{-\pi}^{\pi} \sin nx \cos nx \,dx$

$\sin nx$ is an odd function
$\cos nx$ is an even function
$\therefore \sin nx \cos nx$ is an odd function

$$\therefore \int_{-\pi}^{\pi} \sin nx \cos nx \,dx = 0$$

Exercise 11

Determine whether each of the following function products is odd, even, or neither.

1. $x^2 \cos 3x$
2. $x^3 \sin 2x$
3. $\sin 5x \cos 3x$
4. $x^4 \sin nx$
5. $e^x \cos 4x$
6. $\dfrac{x^2}{2} \cos^2 x$
7. $(3x + 2) \sin 3x$
8. $4x \cos \dfrac{x}{2}$
9. $5 \sin^2 x \cos nx$
10. $(x^2 + 2) \sin nx$

3.7.4 Two important theorems

Theorem 1 If $f(x)$ is defined over the interval $-\pi < x < \pi$ and $f(x)$ is *even*, then the Fourier series has *cosine terms only*. This includes a_0 which may be regarded as $a_n \cos nx$ with $n = 0$.

The Fourier coefficients are given by

$$a_0 = \frac{2}{\pi}\int_0^{\pi} f(x)\,dx$$

$$a_n = \frac{2}{\pi}\int_0^{\pi} f(x)\cos nx\,dx$$

$$b_n = 0 \qquad \text{for } n = 1, 2, 3, \ldots, \text{ in each case}$$

Proof:

(a) Since $f(x)$ is even, $\int_{-\pi}^{0} f(x)\,dx = \int_0^{\pi} f(x)\,dx$.

$$\therefore\ a_0 = \frac{1}{\pi}\int_{-\pi}^{\pi} f(x)\,dx = \frac{2}{\pi}\int_0^{\pi} f(x)\,dx$$

(b) Now $\cos nx$ is also an even function. Therefore the product $f(x) \cos nx$ is a product of two even functions and is therefore itself even.

$$\therefore\ a_n = \frac{1}{\pi}\int_{-\pi}^{\pi} f(x)\cos nx\,dx = \frac{2}{\pi}\int_0^{\pi} f(x)\cos nx\,dx$$

(c) Also, $\sin nx$ is an odd function. Therefore, the product $f(x) \sin nx$ is the product of an even function and an odd function and is therefore itself odd.

$$\therefore\ b_n = \frac{1}{\pi}\int_{-\pi}^{\pi} f(x)\sin nx\,dx = 0$$

The Fourier series for an even function contains cosine terms only (including the constant term a_0).

Knowing this result, there is, of course, no need to waste time going through the working from first principles to find the coefficients b_n of the sine terms, only to find they are all zero.

Example

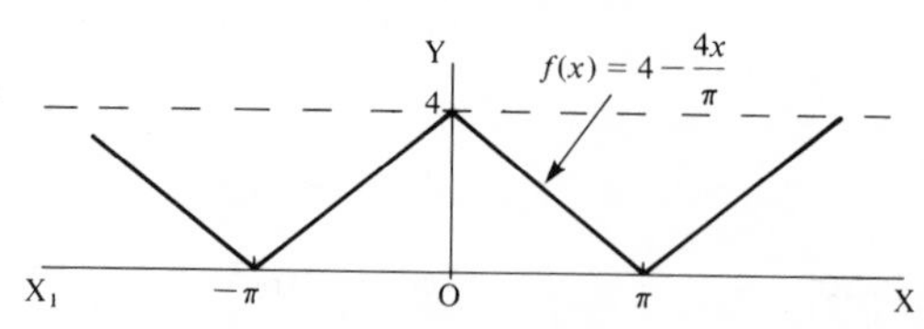

The waveform shown is symmetrical about OY. Therefore, the function is even.

Hence, its Fourier series contains cosines terms only, i.e. $b_n = 0$.

$$\therefore\ f(x) = \tfrac{1}{2}a_0 + \sum_{n=1}^{\infty} a_n \cos nx$$

$$a_0 = \frac{1}{\pi}\int_{-\pi}^{\pi} f(x)\,\mathrm{d}x = \frac{2}{\pi}\int_0^{\pi} f(x)\,\mathrm{d}x$$

Between $x = 0$ and $x = \pi$, $f(x) = 4 - \dfrac{4}{\pi}x$.

$$\therefore\ a_0 = \frac{8}{\pi}\int_0^{\pi}\left(1 - \frac{x}{\pi}\right)\mathrm{d}x$$

$$\therefore\ \frac{\pi a_0}{8} = \int_0^{\pi}\left(1 - \frac{x}{\pi}\right)\mathrm{d}x = \left[x - \frac{x^2}{2\pi}\right]_0^{\pi} = \left\{\pi - \frac{\pi^2}{2\pi}\right\} - \{0\} = \frac{\pi}{2}$$

$$\therefore\ a_0 = \left(\frac{8}{\pi}\right)\left(\frac{\pi}{2}\right) = 4 \qquad\qquad \therefore\ \underline{\tfrac{1}{2}a_0 = 2}$$

$$a_n = \frac{2}{\pi}\int_0^{\pi} f(x)\cos nx\,\mathrm{d}x = \frac{8}{\pi}\int_0^{\pi}\left(1 - \frac{x}{\pi}\right)\cos nx\,\mathrm{d}x$$

$$\therefore\ \frac{\pi a_n}{8} = \int_0^{\pi}\cos nx\,\mathrm{d}x - \frac{1}{\pi}\int_0^{\pi} x\cos nx\,\mathrm{d}x$$

$$= \left[\frac{\sin nx}{n}\right]_0^{\pi} - \frac{1}{\pi}\left\{\left[x\left(\frac{\sin nx}{n}\right)\right]_0^{\pi} - \frac{1}{n}\int_0^{\pi}\sin nx\,\mathrm{d}x\right\}$$

$$= \quad 0 \quad - \frac{1}{\pi}\left\{ \quad 0 \quad - \frac{1}{n}\left[\frac{-\cos nx}{n}\right]_0^{\pi}\right\}$$

$$= \frac{1}{\pi n^2}\{1 - \cos n\pi\} \qquad\qquad \begin{array}{l} n \text{ odd, } \cos n\pi = -1 \\ n \text{ even, } \cos n\pi = 1 \end{array}$$

$$= \frac{2}{\pi n^2} \quad (n \text{ odd}) \qquad \text{and} \quad 0\ (n \text{ even})$$

$$\therefore\ a_n = \frac{16}{\pi^2 n^2} \quad (n \text{ odd}) \qquad \text{and} \quad 0\ (n \text{ even})$$

$$\therefore\ f(x) = 2 + \frac{16}{\pi^2}\left\{\cos x + \frac{\cos 3x}{3^2} + \frac{\cos 5x}{5^2} + \frac{\cos 7x}{7^2} + \ldots\right\}$$

Theorem 2 If $f(x)$ is defined over the interval $-\pi < x < \pi$ and $f(x)$ is *odd*, the Fourier series for $f(x)$ has *sine terms only*.

The coefficients are then given by

$$a_0 = 0; \qquad a_n = 0$$

$$b_n = \frac{2}{\pi}\int_0^{\pi} f(x) \sin nx \, dx$$

Proof:

(a) $a_0 = \frac{1}{\pi}\int_{-\pi}^{\pi} f(x)\, dx.$

Since $f(x)$ is odd,

$$\int_{-\pi}^{0} f(x)\, dx = -\int_0^{\pi} f(x)\, dx \qquad \therefore\ a_0 = 0$$

(b) $a_n = \frac{1}{\pi}\int_{-\pi}^{\pi} f(x) \cos nx\, dx.$

The product $f(x) \cos nx = (\text{odd}) \times (\text{even})$, i.e. (odd).

$$\therefore \int_{-\pi}^{\pi} f(x) \cos nx\, dx = \int_{-\pi}^{\pi} (\text{odd function})\, dx = 0 \qquad \therefore\ a_n = 0$$

(c) $b_n = \frac{1}{\pi}\int_{-\pi}^{\pi} f(x) \sin nx\, dx.$

The product $f(x) \sin nx = (\text{odd}) \times (\text{odd})$, i.e. (even).

$$\therefore \int_{-\pi}^{\pi} f(x) \sin nx\, dx = \int_{-\pi}^{\pi} (\text{even function})\, dx = 2\int_0^{\pi} f(x) \sin nx\, dx$$

$$\therefore\ b_n = \frac{2}{\pi}\int_0^{\pi} f(x) \sin nx\, dx \qquad (n = 1, 2, 3, \ldots)$$

∴ *The Fourier series for an odd function contains sine terms only.*

Example

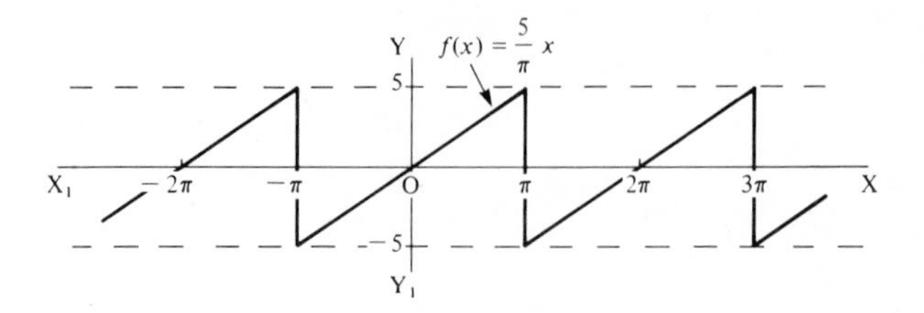

The waveform shown is an odd function, since $f(-x) = -f(x)$ for all x. The Fourier series therefore contains sine terms only, i.e. $a_0 = a_n = 0$.

$$b_n = \frac{1}{\pi}\int_{-\pi}^{\pi} f(x)\sin nx\,dx = \frac{2}{\pi}\int_0^{\pi} \frac{5}{\pi}x\sin nx\,dx$$

$$\therefore\ \frac{\pi^2 b_n}{10} = \int_0^{\pi} x\sin nx\,dx = \left[x\left(\frac{-\cos nx}{n}\right)\right]_0^{\pi} + \frac{1}{n}\int_0^{\pi}\cos nx\,dx$$

$$= \left[\frac{-\pi\cos n\pi}{n} - 0\right]_0^{\pi} + \frac{1}{n}\left[\frac{\sin nx}{n}\right]_0^{\pi}$$

$$= \frac{-\pi\cos n\pi}{n} \qquad \therefore\ b_n = -\frac{10}{n\pi}\cos n\pi \quad (n = 1, 2, 3, \ldots)$$

When n is odd, $\cos n\pi = -1$
When n is even, $\cos n\pi = 1$

$$b_n = \frac{10}{\pi}\left\{-\frac{\cos n\pi}{n}\right\}$$

$$\therefore\ f(x) = \frac{10}{\pi}\left\{\sin x - \frac{\sin 2x}{2} + \frac{\sin 3x}{3} - \frac{\sin 4x}{4} + \ldots\right\}$$

Summary

$$f(x) = \tfrac{1}{2}a_0 + \sum_{n=1}^{\infty}\{a_n\cos nx + b_n\sin nx\}$$

$$a_0 = \frac{1}{\pi}\int_{-\pi}^{\pi} f(x)\,dx$$

$$a_n = \frac{1}{\pi}\int_{-\pi}^{\pi} f(x)\cos nx\,dx$$

$$b_n = \frac{1}{\pi}\int_{-\pi}^{\pi} f(x)\sin nx\,dx$$

(a) If $f(x)$ is an *even function*, i.e. symmetrical about the y-axis, the Fourier series contains *cosine terms only* (including a_0). Then

$$a_0 = \frac{2}{\pi}\int_0^{\pi} f(x)\,\mathrm{d}x$$

$$a_n = \frac{2}{\pi}\int_0^{\pi} f(x)\cos nx\,\mathrm{d}x$$

$$b_n = 0$$

(b) If $f(x)$ is an *odd function*, i.e. symmetrical about the origin, the Fourier series contains *sine terms only*. Then

$$a_0 = 0$$

$$a_n = 0$$

$$b_n = \frac{2}{\pi}\int_0^{\pi} f(x)\sin nx\,\mathrm{d}x$$

Exercise 12

For each of the following functions, determine the Fourier series to represent the function, using your knowledge of odd and even functions.

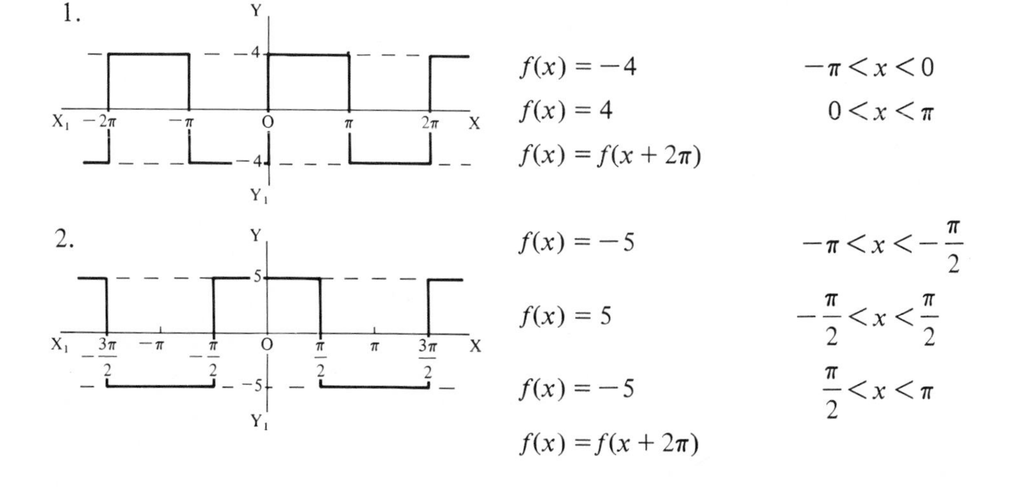

1. $f(x) = -4 \qquad -\pi < x < 0$

$f(x) = 4 \qquad 0 < x < \pi$

$f(x) = f(x + 2\pi)$

2. $f(x) = -5 \qquad -\pi < x < -\frac{\pi}{2}$

$f(x) = 5 \qquad -\frac{\pi}{2} < x < \frac{\pi}{2}$

$f(x) = -5 \qquad \frac{\pi}{2} < x < \pi$

$f(x) = f(x + 2\pi)$

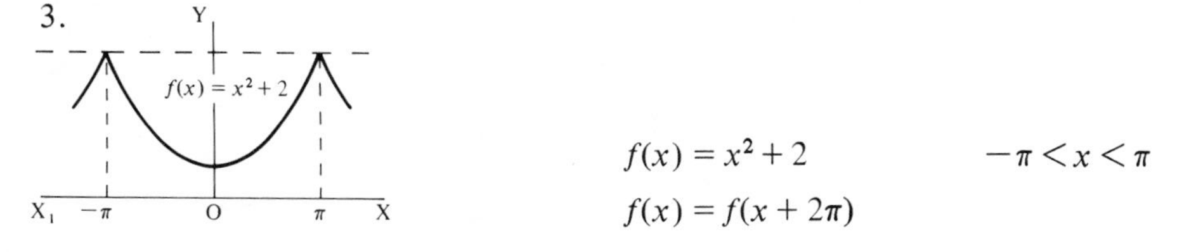

3. $f(x) = x^2 + 2 \qquad -\pi < x < \pi$

$f(x) = f(x + 2\pi)$

4.

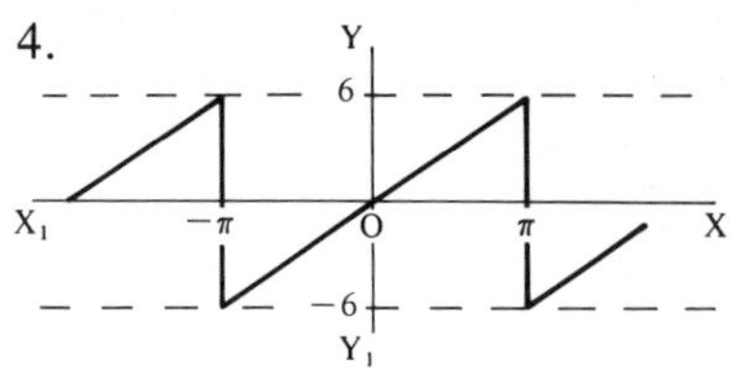

$$f(x) = \frac{6x}{\pi} \qquad -\pi < x < \pi$$

$$f(x) = f(x + 2\pi)$$

5.

$$f(x) = -A \qquad -\pi < x < -\frac{\pi}{2}$$

$$f(x) = 0 \qquad -\frac{\pi}{2} < x < \frac{\pi}{2}$$

$$f(x) = A \qquad \frac{\pi}{2} < x < \pi$$

$$f(x) = f(x + 2\pi)$$

6.

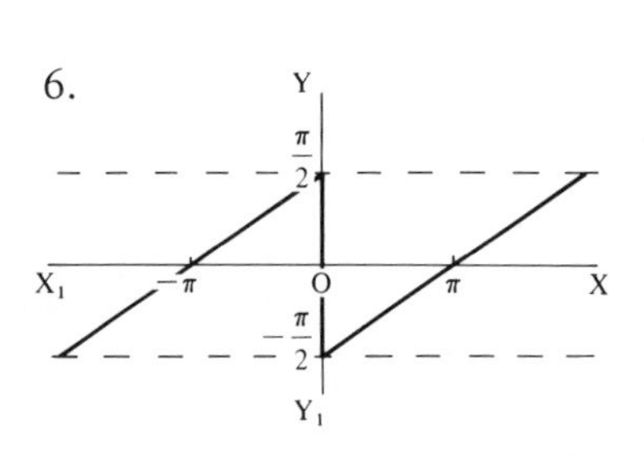

$$f(x) = \frac{x}{2} + \frac{\pi}{2} \qquad -\pi < x < 0$$

$$f(x) = \frac{x}{2} - \frac{\pi}{2} \qquad 0 < x < \pi$$

$$f(x) = f(x + 2\pi)$$

3.8 REVISION SUMMARY

1. Fourier series

Function $f(x)$ of period 2π.

$$f(x) = \tfrac{1}{2}a_0 + a_1 \cos x + a_2 \cos 2x + a_3 \cos 3x + \ldots$$
$$+ b_1 \sin x + b_2 \sin 2x + b_3 \sin 3x + \ldots$$

$$f(x) = \tfrac{1}{2}a_0 + \sum_{n=1}^{\infty} \{a_n \cos nx + b_n \sin nx\}$$

2. Dirichlet conditions

A function can be represented as a Fourier series if certain conditions are satisfied:

(a) $f(x)$ must be defined and single-valued throughout the cycle;

(b) $f(x)$ must be continuous or have a finite number of finite discontinuities in the periodic interval;

(c) $f(x)$ and $f'(x)$ must be piecewise continuous in the periodic interval.

3. Fourier coefficients

$$f(x) = \tfrac{1}{2}a_0 + \sum_{n=1}^{\infty}\{a_n \cos nx + b_n \sin nx\}$$

$$a_0 = \frac{1}{\pi}\int_{-\pi}^{\pi} f(x)\,dx \qquad = 2 \times \text{mean value of } f(x) \text{ over a period}$$

$$a_n = \frac{1}{\pi}\int_{-\pi}^{\pi} f(x)\cos nx\,dx = 2 \times \text{mean value of } f(x)\cos nx \text{ over a period}$$

$$b_n = \frac{1}{\pi}\int_{-\pi}^{\pi} f(x)\sin nx\,dx = 2 \times \text{mean value of } f(x)\sin nx \text{ over a period}$$

$(n = 1, 2, 3, \ldots)$

4. Fourier series at a point of discontinuity

If $x = x_1$ is a point of continuity, the series converges to the value of $f(x_1)$.

If $x = x_1$ is a point of finite discontinuity, the series converges to the value of $\dfrac{f(x_1 - 0) + f(x_1 + 0)}{2}$.

5. Odd and even functions

Even function: $f(-x) = f(x)$ i.e. symmetrical about the y-axis.

Odd function: $f(-x) = -f(x)$ i.e. symmetrical about the origin.

Products of odd and even functions: similar to 'rules of signs'.

If $f(x)$ and $g(x)$ are *both even*, $f(x)\,g(x)$ is *even*.

If $f(x)$ and $g(x)$ are *both odd*, $f(x)\,g(x)$ is *even*.

If one is *even* and the other *odd*, $f(x)\,g(x)$ is *odd*.

6. Fourier series of odd and even functions

(a) If $f(x)$ is an *even function*, i.e. symmetrical about the y-axis, the Fourier series contains *cosine terms only* (including $\tfrac{1}{2}a_0$). Then

$$a_0 = \frac{2}{\pi}\int_{0}^{\pi} f(x)\,dx$$

$$a_n = \frac{2}{\pi}\int_{0}^{\pi} f(x)\cos nx\,dx$$

$$b_n = 0 \qquad (n = 1, 2, 3, \ldots)$$

(b) If $f(x)$ is an *odd function*, i.e. symmetrical about the origin, the Fourier series contains *sine terms only*. Then

$$a_0 = 0$$

$$a_n = 0$$

$$b_n = \frac{2}{\pi}\int_0^{\pi} f(x) \sin nx \, dx \qquad (n = 1, 2, 3, \ldots)$$

Chapter 4

HALF-RANGE SERIES

4.1 FUNCTIONS DEFINED OVER HALF A PERIOD

If a function is defined over the range $-\pi$ to π, the Fourier series representing the function will, in general, contain both sine terms and cosine terms.

$$f(x) = \tfrac{1}{2}a_0 + \sum_{n=1}^{\infty}\{a_n \cos nx + b_n \sin nx\}$$

where
$$a_0 = \frac{1}{\pi}\int_{-\pi}^{\pi} f(x)\,dx$$

$$a_n = \frac{1}{\pi}\int_{-\pi}^{\pi} f(x)\cos nx\,dx$$

$$b_n = \frac{1}{\pi}\int_{-\pi}^{\pi} f(x)\sin nx\,dx$$

We have also seen that, in certain circumstances, the series will contain only cosine terms (even function), or only sine terms (odd function).

Sometimes a function of period 2π is defined over a range of 0 to π, instead of $-\pi$ to π, and in this case we have a certain amount of choice of how to proceed.

For example, let us take the case when the function is defined by

$$f(x) = 1 + x \qquad 0 < x < \pi$$

and
$$f(x) = f(x + 2\pi)$$

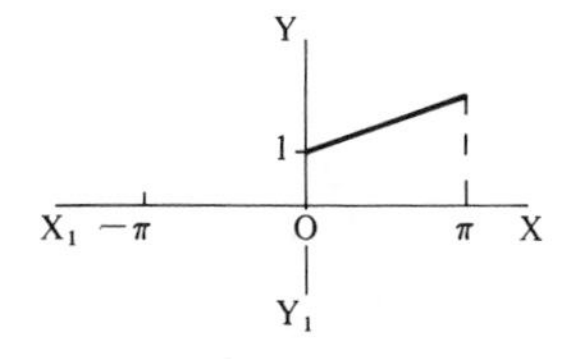

We are not told what happens between $-\pi$ and 0.

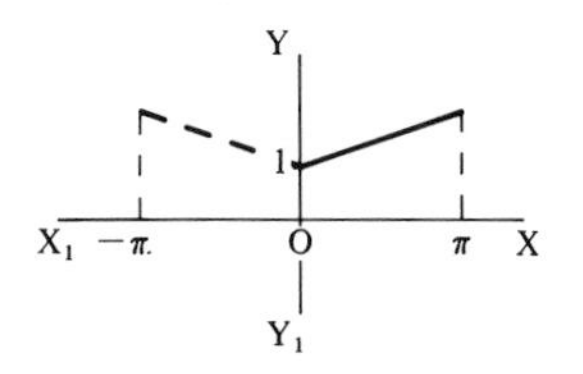

If by chance, the waveform were as shown here, then the function would clearly be an even function, being symmetrical about the y-axis and the series would contain only cosine terms, including a_0.

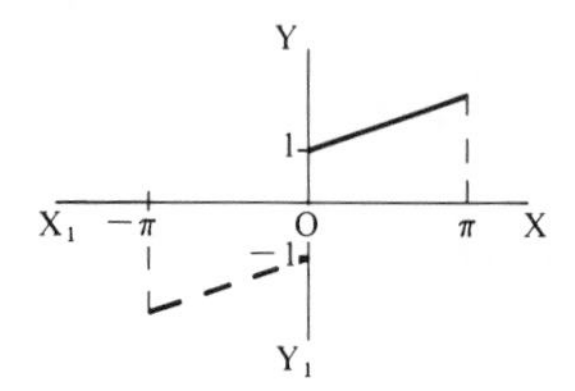

If, however, the waveform were as shown here, then the function would now be an odd function, being symmetrical about the origin, and the series would contain only sine terms.

Of course, the waveform between $-\pi$ and 0 might, in fact, be something quite different, e.g.

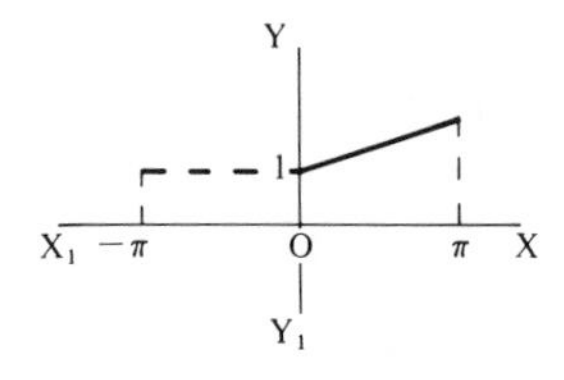

In this case, the function would be neither odd nor even and the resulting series would contain both sine and cosine terms.

Whichever arrangement we choose to adopt, the resulting Fourier series will represent $f(x)$ in the range 0 to π only. It will not necessarily represent the function outside this range, since we have no information of how the function really behaves outside these limits.

Such a series, for obvious reason, is called a *half-range series* for $f(x)$, and we normally elect to represent the function either as a *sine series* or as a *cosine series*.

Example 1

Express the function defined by

$$f(x) = x \qquad 0 < x < \pi$$

$$f(x) = f(x + 2\pi)$$

(a) as a half-range sine series

(b) as a half-range cosine series.

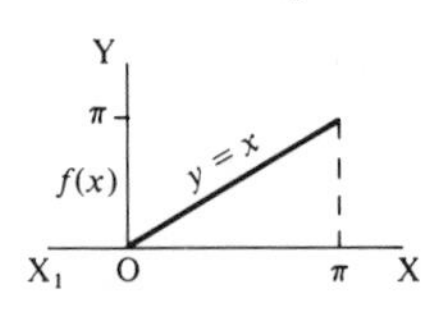

The function is defined only between $x = 0$ and $x = \pi$.

(a) **Half-range sine series**

To obtain a sine series, we select the waveform between $-\pi$ and 0 to give an odd function, i.e. the waveform must be symmetrical about the origin.

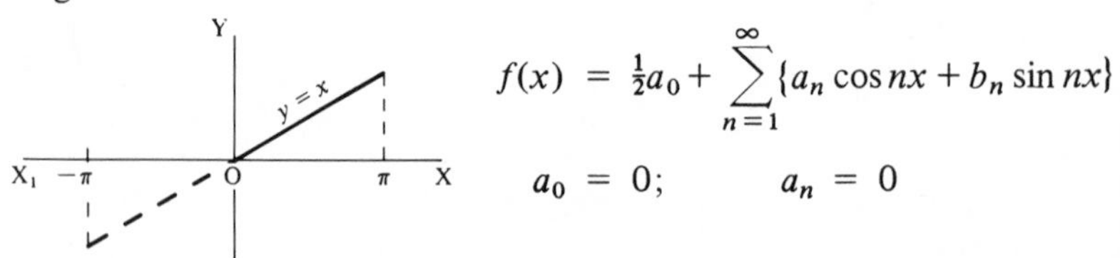

$$f(x) = \tfrac{1}{2}a_0 + \sum_{n=1}^{\infty}\{a_n \cos nx + b_n \sin nx\}$$

$$a_0 = 0; \qquad a_n = 0$$

$$b_n = \frac{1}{\pi}\int_{-\pi}^{\pi} f(x)\sin nx\,\mathrm{d}x$$

$$= \frac{2}{\pi}\int_{0}^{\pi} x\sin nx\,\mathrm{d}x$$

$$\therefore \frac{\pi b_n}{2} = \left[x\left(\frac{-\cos nx}{n}\right)\right]_0^{\pi} + \frac{1}{n}\int_0^{\pi}\cos nx\,\mathrm{d}x$$

$$= \left[-\frac{x\cos nx}{n}\right]_0^{\pi} + \frac{1}{n}\left[\frac{\sin nx}{n}\right]_0^{\pi}$$

$$= \left[-\frac{\pi\cos n\pi}{n} + 0\right] + \frac{1}{n}[0-0] = -\frac{\pi\cos n\pi}{n}$$

$$\therefore b_n = -\frac{2}{n}\cos n\pi \qquad \cos n\pi = -1\ (n\text{ odd})$$
$$= 1\quad (n\text{ even})$$

$$\therefore b_n = \frac{2}{n}\quad (n\text{ odd}) \qquad\text{and}\qquad b_n = -\frac{2}{n}\quad (n\text{ even})$$

$$\therefore f(x) = 2\left\{\sin x - \frac{1}{2}\sin 2x + \frac{1}{3}\sin 3x - \frac{1}{4}\sin 4x + \ldots\right\}$$

(b) **Half-range cosine series**

To obtain a cosine series, we must arrange the waveform between $-\pi$ and 0 to be symmetrical about the y-axis, i.e. to represent an even function.

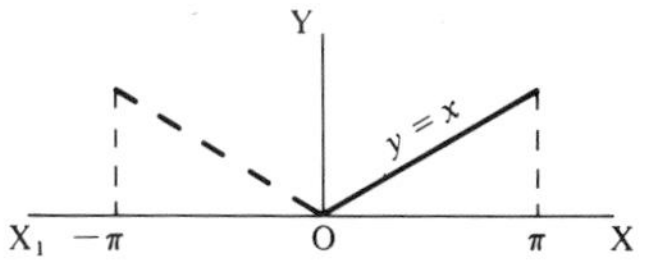

$$f(x) = \tfrac{1}{2}a_0 + \sum_{n=1}^{\infty}\{a_n\cos nx + b_n\sin nx\}$$

$$a_0 = \frac{1}{\pi}\int_{-\pi}^{\pi} f(x)\,\mathrm{d}x = \frac{2}{\pi}\int_0^{\pi} f(x)\,\mathrm{d}x$$

$$a_n = \frac{1}{\pi}\int_{-\pi}^{\pi} f(x)\cos nx\,\mathrm{d}x$$

$$= \frac{2}{\pi}\int_0^{\pi} f(x)\cos nx\,\mathrm{d}x$$

$$b_n = 0$$

$$\therefore \frac{\pi a_0}{2} = \int_0^{\pi} x\,\mathrm{d}x = \left[\frac{x^2}{2}\right]_0^{\pi} = \frac{\pi^2}{2} \qquad \therefore a_0 = \pi$$

$$\frac{\pi a_n}{2} = \int_0^\pi x \cos nx\,dx = \left[x\left(\frac{\sin nx}{n}\right)\right]_0^\pi - \frac{1}{n}\int_0^\pi \sin nx\,dx$$

$$= \left[\frac{\pi \sin n\pi}{n} - 0\right] - \frac{1}{n}\left[\frac{-\cos nx}{n}\right]_0^\pi$$

$$= [\quad 0 - 0 \quad] + \frac{1}{n^2}[\cos n\pi - 1]$$

$$\therefore a_n = \frac{2}{\pi n^2}\{\cos n\pi - 1\} \qquad \cos n\pi = -1\ (n \text{ odd}) \quad = 1\ (n \text{ even})$$

$$\therefore a_n = \frac{-4}{\pi n^2}\ (n \text{ odd}) \quad \text{and} \quad a_n = 0\ (n \text{ even})$$

$$\therefore f(x) = \frac{\pi}{2} - \frac{4}{\pi}\left\{\cos x + \frac{1}{3^2}\cos 3x + \frac{1}{5^2}\cos 5x + \dots\right\}$$

Example 2

Express the function

$$f(x) = 0 \qquad 0 < x < \frac{\pi}{2}$$

$$f(x) = 1 \qquad \frac{\pi}{2} < x < \pi$$

$$f(x) = f(x + 2\pi)$$

(a) as a half-range sine series

(b) as a half-range cosine series.

(a) **Half-range sine series**

To obtain a sine series, we select the part of the waveform between $-\pi$ and 0 to represent an odd function, symmetrical about the origin.

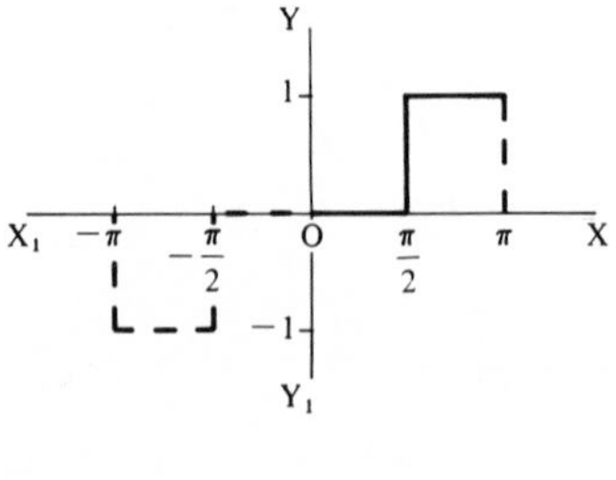

Then $a_0 = 0; \qquad a_n = 0$

$$f(x) = \sum_{n=1}^{\infty} b_n \sin nx$$

where $b_n = \dfrac{2}{\pi}\displaystyle\int_0^\pi f(x) \sin nx\,dx$

$$\therefore \quad \frac{\pi b_n}{2} = \int_0^{\pi/2} 0 \sin nx \, dx + \int_{\pi/2}^{\pi} 1 \sin nx \, dx$$

$$= \left[\frac{-\cos nx}{n}\right]_{\pi/2}^{\pi} = \frac{1}{n}\left\{-\cos n\pi + \cos n\frac{\pi}{2}\right\}$$

$$\therefore \quad b_n = \frac{2}{n\pi}\left\{-\cos n\pi + \cos n\frac{\pi}{2}\right\}$$

Putting $n = 1, 2, 3, \ldots$, gives

b_1	b_2	b_3	b_4	b_5
$\frac{2}{\pi}$	$-\frac{2}{\pi}$	$\frac{2}{3\pi}$	0	$\frac{2}{5\pi}$

$$\therefore \quad f(x) = \frac{2}{\pi}\left\{\sin x - \sin 2x + \frac{1}{3}\sin 3x + \frac{1}{5}\sin 5x + \ldots\right\}$$

(b) **Half-range cosine series**

To obtain a cosine series, we draw the waveform between $-\pi$ and 0 so that the complete trace is symmetrical about the y-axis and therefore represents an even function.

There are no sine terms in the series,

i.e. $\quad b_n = 0$

$$f(x) = \tfrac{1}{2}a_0 + \sum_{n=1}^{\infty} a_n \cos nx$$

$$a_0 = \frac{1}{\pi}\int_{-\pi}^{\pi} f(x)\, dx = \frac{2}{\pi}\int_0^{\pi} f(x)\, dx$$

$$= \frac{2}{\pi}\int_{\pi/2}^{\pi} 1\, dx = \frac{2}{\pi}\Big[x\Big]_{\pi/2}^{\pi} = \frac{2}{\pi}\left\{\pi - \frac{\pi}{2}\right\} = \left(\frac{2}{\pi}\right)\left(\frac{\pi}{2}\right) \qquad \therefore \quad a_0 = 1$$

$$a_n = \frac{1}{\pi}\int_{-\pi}^{\pi} f(x) \cos nx \, dx = \frac{2}{\pi}\int_0^{\pi} f(x) \cos nx \, dx$$

$$= \frac{2}{\pi}\int_{\pi/2}^{\pi} 1 \cos nx = \frac{2}{\pi}\left[\frac{\sin nx}{n}\right]_{\pi/2}^{\pi}$$

$$\therefore \quad a_n = \frac{2}{n\pi}\left\{\sin n\pi - \sin n\frac{\pi}{2}\right\} = -\frac{2}{n\pi}\sin n\frac{\pi}{2}$$

Putting $n = 1, 2, 3, \ldots,$ gives

a_1	a_2	a_3	a_4	a_5
$-\frac{2}{\pi}$	0	$\frac{2}{3\pi}$	0	$-\frac{2}{5\pi}$

$$\therefore\ f(x) = \frac{1}{2} - \frac{2}{\pi}\left\{\cos x - \frac{1}{3}\cos 3x + \frac{1}{5}\cos 5x - \frac{1}{7}\cos 7x + \ldots\right\}$$

Exercise 13

Express each of the following functions as a half-range sine series or a half-range cosine series, as indicated.

1. $f(x) = 1 \qquad 0 < x < \pi \qquad$ (sine series)
2. $f(x) = 2x \qquad 0 < x < \pi \qquad$ (cosine series)
3. $f(x) = 3 \qquad 0 < x < \frac{\pi}{2}$
   $f(x) = 0 \qquad \frac{\pi}{2} < x < \pi \qquad$ (sine series)
4. $f(x) = x \qquad 0 < x < \frac{\pi}{2}$
   $f(x) = \frac{\pi}{2} \qquad \frac{\pi}{2} < x < \pi \qquad$ (cosine series)
5. $f(x) = 1 \qquad 0 < x < \frac{\pi}{2}$
   $f(x) = -1 \qquad \frac{\pi}{2} < x < \pi \qquad$ (cosine series)
6. $f(x) = x \qquad 0 < x < \frac{\pi}{2}$
   $f(x) = 0 \qquad \frac{\pi}{2} < x < \pi \qquad$ (sine series)
7. $f(x) = 4 \qquad 0 < x < \frac{\pi}{2}$
   $f(x) = -4 \qquad \frac{\pi}{2} < x < \pi \qquad$ (sine series)

8. $f(x) = 0 \qquad 0 < x < \frac{\pi}{2}$

$f(x) = 1 \qquad \frac{\pi}{2} < x < \pi$ (cosine series)

9. $f(x) = 1 + x \qquad 0 < x < \pi$ (cosine series)

10. $f(x) = x^2 \qquad 0 < x < \pi$ (sine series)

4.2 SERIES CONTAINING ONLY ODD HARMONICS OR ONLY EVEN HARMONICS

$$f(x) = \tfrac{1}{2}a_0 + \sum_{n=1}^{\infty} \{a_n \cos nx + b_n \sin nx\}$$

$$= \tfrac{1}{2}a_0 + a_1 \cos x + a_2 \cos 2x + a_3 \cos 3x + \ldots$$
$$+ b_1 \sin x + b_2 \sin 2x + b_3 \sin 3x + \ldots$$

For $f(x + \pi)$, the variable x in each term is replaced by $(x + \pi)$.

$$\cos nx \text{ becomes } \cos n(x + \pi), \text{ i.e. } \cos(nx + n\pi)$$

$$\therefore \cos n(x + \pi) = \cos nx \cos n\pi - \sin nx \sin n\pi$$

For $n = 1, 2, 3, \ldots,\ \sin n\pi = 0$

$$\therefore \cos n(x + \pi) = \cos nx \cos n\pi$$

and $\cos n\pi = 1$ for $n = 0, 2, 4, \ldots$ i.e. n even

$= -1$ for $n = 1, 3, 5, \ldots$ i.e. n odd

Similarly,

$$\sin n(x + \pi) = \sin(nx + n\pi) = \sin nx \cos n\pi + \cos nx \sin n\pi$$

For $n = 1, 2, 3, \ldots,\ \sin n\pi = 0$

$$\therefore \sin n(x + \pi) = \sin nx \cos n\pi$$

and $\cos n\pi = 1$ for $n = 0, 2, 4, \ldots$ i.e. n even

$= -1$ for $n = 1, 3, 5, \ldots$ i.e. n odd

$$\therefore f(x + \pi) = \tfrac{1}{2}a_0 - a_1 \cos x + a_2 \cos 2x - a_3 \cos 3x + \ldots$$
$$- b_1 \sin x + b_2 \sin 2x - b_3 \sin 3x + \ldots$$

Now, if $f(x) = f(x + \pi)$

$$\tfrac{1}{2}a_0 + a_1 \cos x + a_2 \cos 2x + a_3 \cos 3x + \ldots + b_1 \sin x + b_2 \sin 2x + b_3 \sin 3x \ldots$$
$$= \tfrac{1}{2}a_0 - a_1 \cos x + a_2 \cos 2x - a_3 \cos 3x + \ldots - b_1 \sin 2x + b_2 \sin 2x - b_3 \sin 3x \ldots$$

For these two series to be equal, the odd harmonics, which differ in sign, must all be zero.

$$\therefore\ f(x) = f(x+\pi) = \tfrac{1}{2}a_0 + a_2\cos 2x + a_4\cos 4x + a_6\cos 6x + \ldots + b_2\sin 2x + b_4\sin 4x + b_6\sin 6x + \ldots$$

i.e. *If $f(x) = f(x+\pi)$, the resulting Fourier series contains even harmonics only.*

Similarly, if $f(x) = -f(x+\pi)$, then

$$\tfrac{1}{2}a_0 + a_1\cos x + a_2\cos 2x + a_3\cos 3x + \ldots + b_1\sin x + b_2\sin 2x + b_3\sin 3x \ldots$$
$$= -\tfrac{1}{2}a_0 + a_1\cos x - a_2\cos 2x + a_3\cos 3x + \ldots + b_1\sin x - b_2\sin 2x + b_3\sin 3x \ldots$$

For these two series to be equal, the even harmonics (including a_0) must be zero.

$$\therefore\ f(x) = -f(x+\pi) = a_1\cos x + a_3\cos 3x + a_5\cos 5x + \ldots + b_1\sin x + b_3\sin 3x + b_5\sin 5x + \ldots$$

i.e. *If $f(x) = -f(x+\pi)$, the resulting Fourier series contains odd harmonics only.*

Example 1

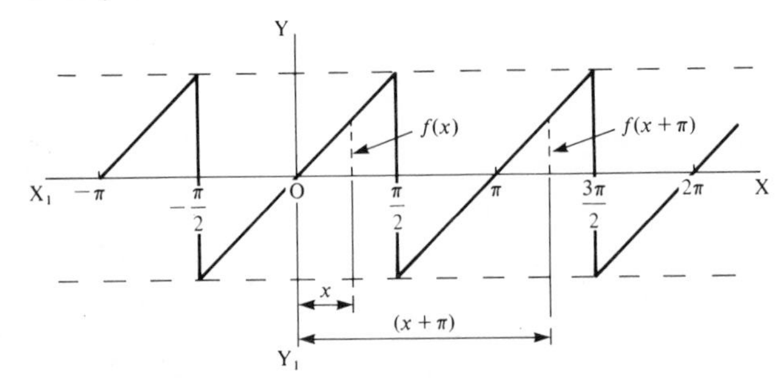

We see that

$$f(x) = f(x+\pi)$$

$\therefore$ even harmonics only.

In fact, if we calculate the coefficients, we obtain

$$f(x) = \sin 2x - \frac{1}{2}\sin 4x + \frac{1}{3}\sin 6x - \frac{1}{4}\sin 8x + \ldots$$

Example 2

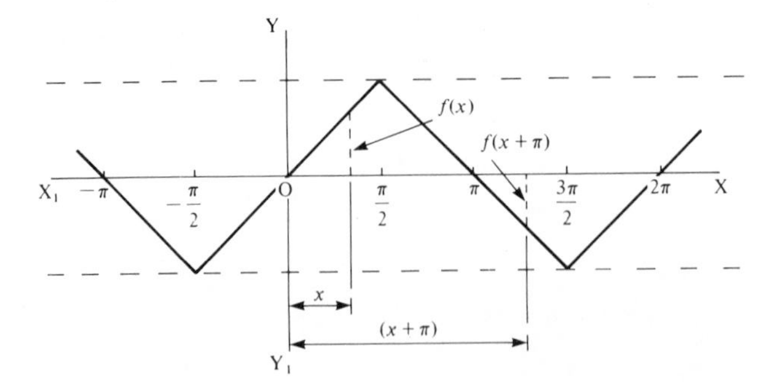

In this case,

$$f(x) = -f(x+\pi)$$

$\therefore$ odd harmonics only.

The relevant Fourier series is, in fact

$$f(x) = \frac{4}{\pi}\left\{\sin x - \frac{1}{9}\sin 3x + \frac{1}{25}\sin 5x - \frac{1}{49}\sin 7x + \ldots\right\}$$

Exercise 14

In each of the following cases, comment on the nature of the terms in the Fourier series.

1.

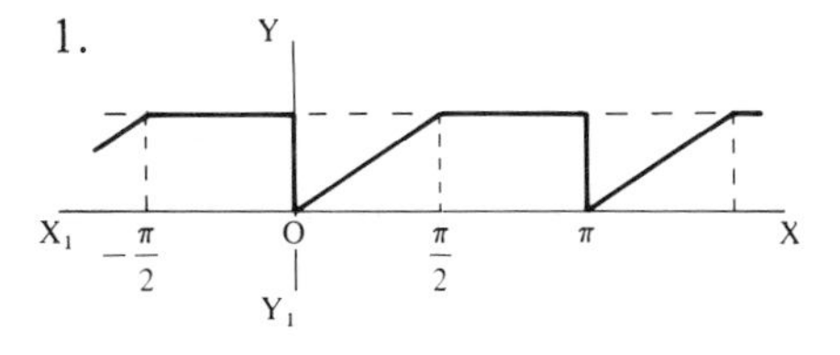

2.

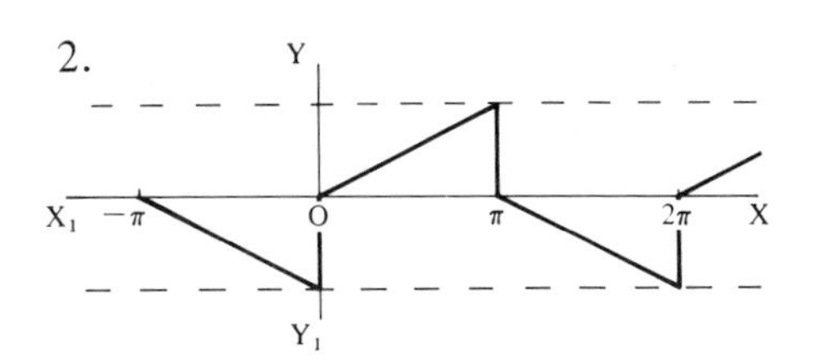

3.

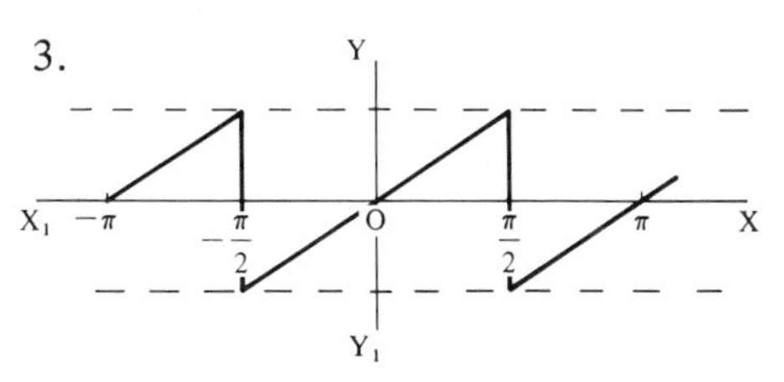

4.

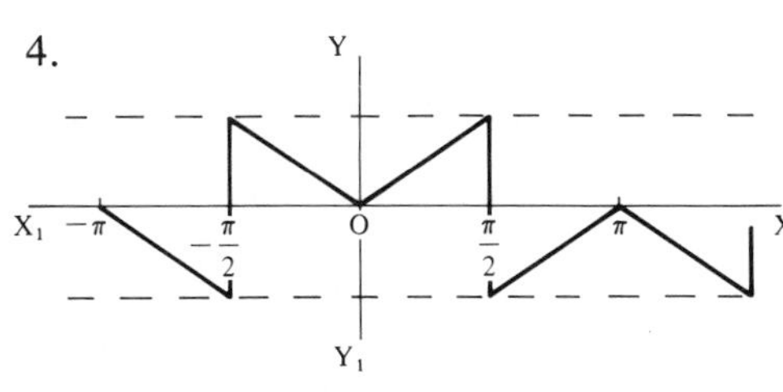

5.

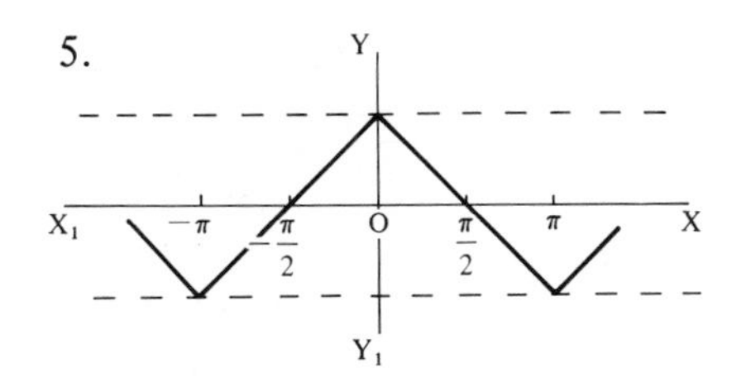

6.

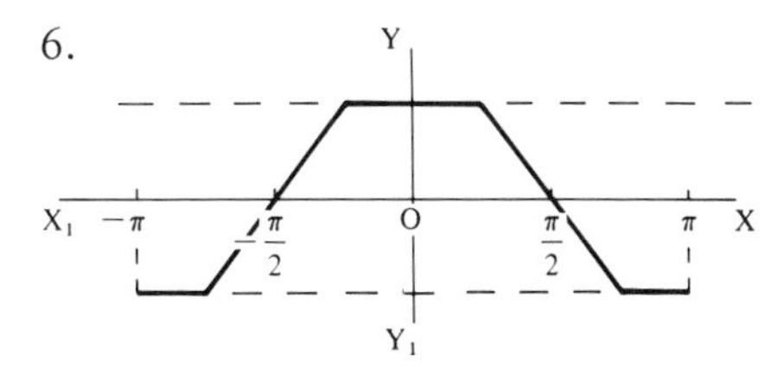

7.

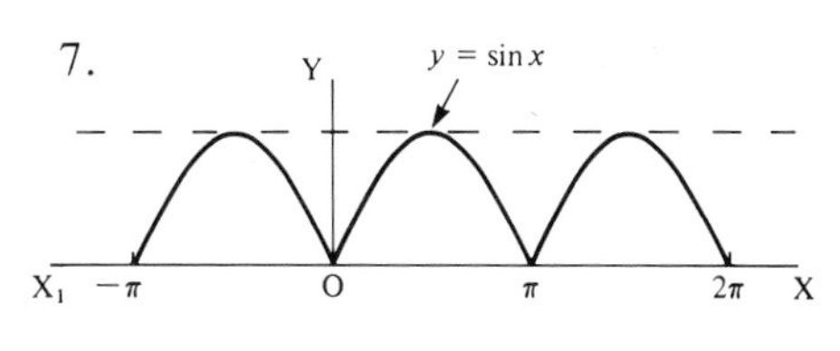

8.

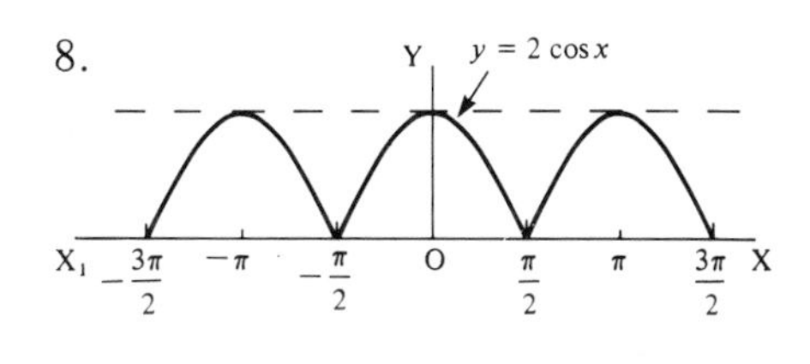

9.

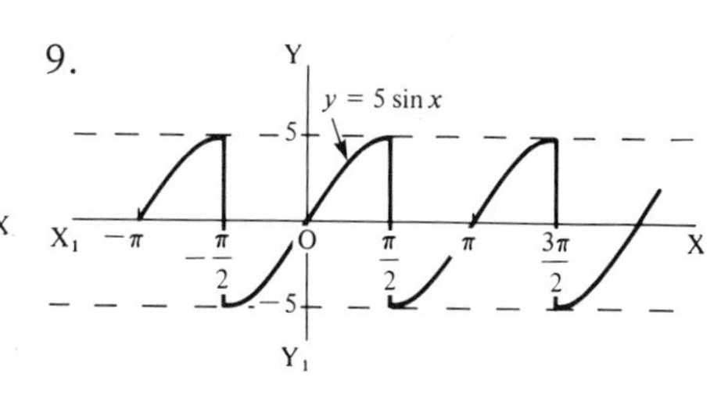

10.

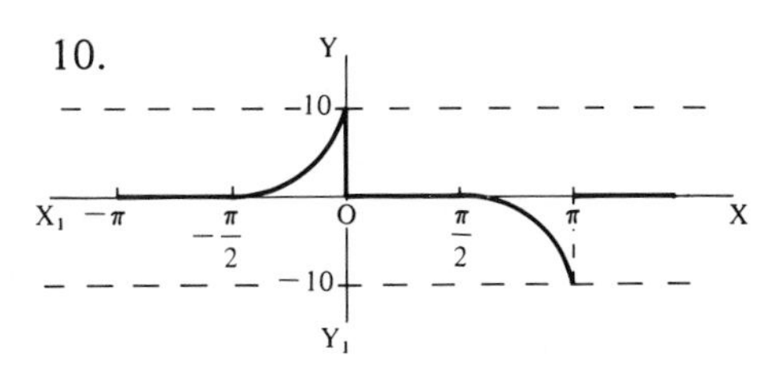

4.3 SIGNIFICANCE OF THE CONSTANT TERM $\frac{1}{2}a_0$

Consider the waveform shown. It represents a periodic function of period 2π.

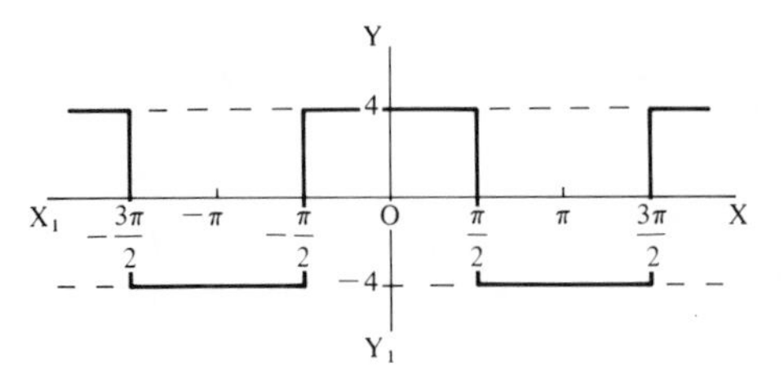

(a) It depicts an even function (symmetrical about the y-axis) $\therefore$ no sine terms.

(b) Average value over one cycle = 0,
$\therefore a_0 = 0$
$\therefore$ constant term $\frac{1}{2}a_0 = 0$.

(c) $f(x) = -f(x+\pi)$,
$\therefore$ odd harmonics only.

$$a_n = \frac{2}{\pi}\int_0^{\pi} f(x)\cos nx\,dx$$

$$\frac{\pi a_n}{2} = \int_0^{\pi/2} 4\cos nx\,dx - \int_{\pi/2}^{\pi} 4\cos nx\,dx$$

$$\therefore \frac{\pi a_n}{8} = \left[\frac{\sin nx}{n}\right]_0^{\pi/2} - \left[\frac{\sin nx}{n}\right]_{\pi/2}^{\pi}$$

$$= \frac{1}{n}\left\{\sin\frac{n\pi}{2} - 0 - \sin n\pi + \sin\frac{n\pi}{2}\right\} \qquad n = 1, 2, 3, \ldots$$

$$= \frac{1}{n}2\sin\frac{n\pi}{2} \qquad \therefore a_n = \frac{16}{n\pi}\sin\frac{n\pi}{2}$$

n	1	3	5	7	9
$\sin\frac{n\pi}{2}$	1	−1	1	−1	1

$$\therefore f(x) = \frac{16}{\pi}\left\{\cos x - \frac{1}{3}\cos 3x + \frac{1}{5}\cos 5x - \frac{1}{7}\cos 7x + \ldots\right\}$$

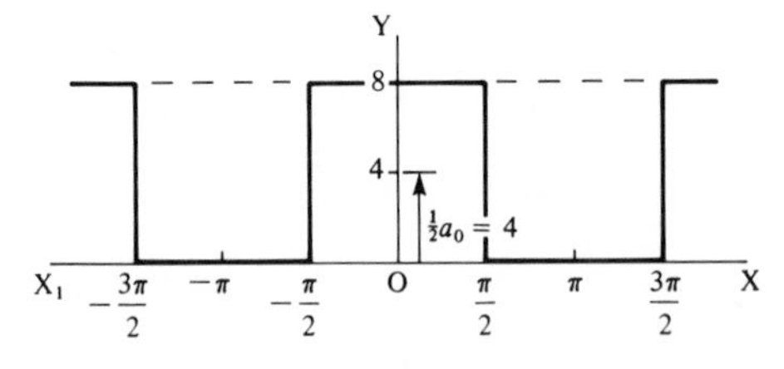

If we now consider a slightly different example, we see that the waveform is symmetrical about the y-axis and therefore represents an even function. There will therefore be no sine terms.

$$a_0 = \frac{2}{\pi}\int_0^{\pi} f(x)\,dx = \frac{2}{\pi}\int_0^{\pi/2} 8\,dx = \frac{16}{\pi}\Big[x\Big]_0^{\pi/2} = \left(\frac{16}{\pi}\right)\left(\frac{\pi}{2}\right) = 8 \quad \therefore \ \underline{\tfrac{1}{2}a_0 = 4}$$

$$a_n = \frac{2}{\pi}\int_0^{\pi} f(x)\cos nx\,dx = \frac{2}{\pi}\int_0^{\pi/2} 8\cos nx\,dx = \frac{16}{\pi}\left[\frac{\sin nx}{n}\right]_0^{\pi/2}$$

$$= \frac{16}{n\pi}\left\{\sin\frac{n\pi}{2}\right\} \qquad \therefore \ \underline{a_n = \frac{16}{n\pi}\sin\frac{n\pi}{2}}$$

n	1	2	3	4	5	6	7
$\sin\frac{n\pi}{2}$	1	0	−1	0	1	0	−1

$$\therefore \ \underline{f(x) = 4 + \frac{16}{\pi}\left\{\cos x - \frac{1}{3}\cos 3x + \frac{1}{5}\cos 5x - \frac{1}{7}\cos 7x + \ldots\right\}}$$

Notice that the right hand side expression is that of the previous example with the addition of the constant term $\frac{1}{2}a_0$, i.e. 4. That is, the whole waveform has been lifted through a distance given by the value of $\frac{1}{2}a_0$.

Similarly, for the waveform positioned thus

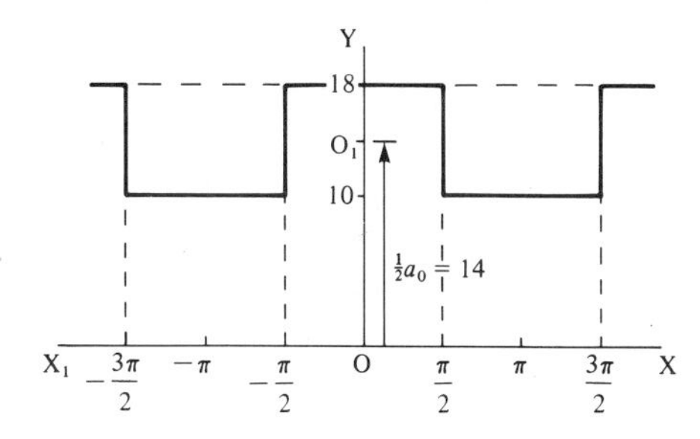

the original origin (O_1) has been lifted 14 units and the series is

$$f(x) = 14 + \frac{16}{\pi}\left\{\cos x - \frac{1}{3}\cos 3x \ldots\right\}$$

We can often with advantage apply this principle in reverse to save unnecessary calculations.

Example

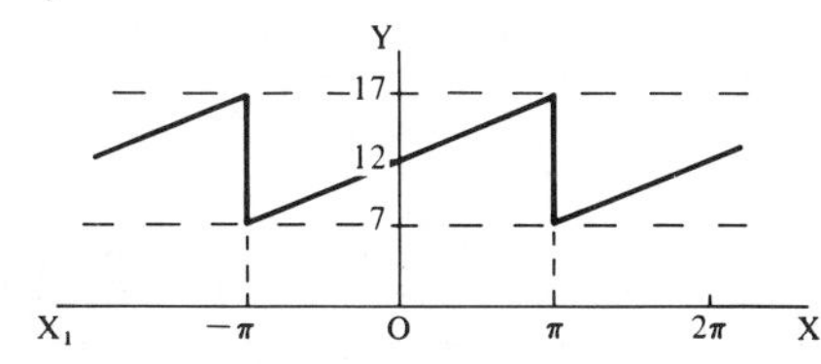

Express the waveform shown as a Fourier series.

This is not an even function or an odd function and therefore calculation of the usual coefficients, a_0, a_n, b_n, would be necessary.

However, if we consider the waveform lowered by 12 units to its elementary position, it becomes

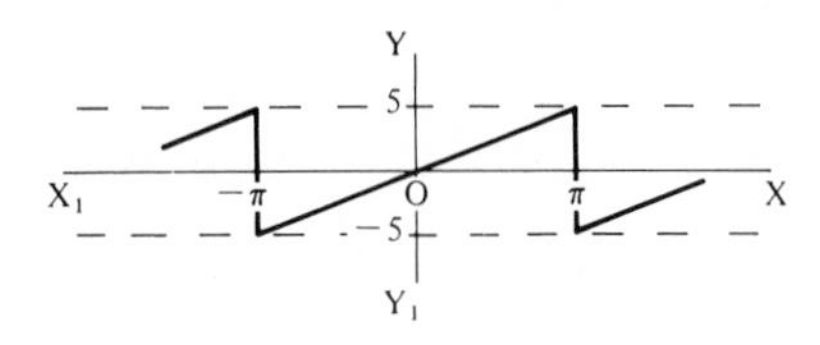

This depicts an odd function. Therefore sine terms only, i.e. $a_0 = 0$ and $a_n = 0$. Treating it as a half-range sine series, we have

$$b_n = \frac{2}{\pi}\int_0^{\pi} f(x)\sin nx\,dx$$

$$\frac{\pi b_n}{2} = \int_0^{\pi} \frac{5}{\pi}x \sin nx\,dx$$

$$\frac{\pi^2 b_n}{10} = \left[x\left(\frac{-\cos nx}{n}\right)\right]_0^{\pi} + \frac{1}{n}\int_0^{\pi}\cos nx\,dx$$

$$= \left[\frac{-x\cos nx}{n}\right]_0^{\pi} + \frac{1}{n}\left[\frac{\sin nx}{n}\right]_0^{\pi}$$

$$= \frac{1}{n}\left\{-\pi\cos n\pi + 0\right\} + \frac{1}{n^2}\left\{\sin n\pi - 0\right\} \qquad n = 1, 2, 3, \ldots$$

$$= -\frac{\pi}{n}\cos n\pi \qquad \therefore\ b_n = -\frac{10}{n\pi}\cos n\pi$$

n	1	2	3	4	5	6	7
$\cos n\pi$	−1	1	−1	1	−1	1	−1

$$\therefore\ f(x) = \frac{10}{\pi}\left\{\sin x - \frac{1}{2}\sin 2x + \frac{1}{3}\sin 3x - \frac{1}{4}\sin 4x + \ldots\right\}$$

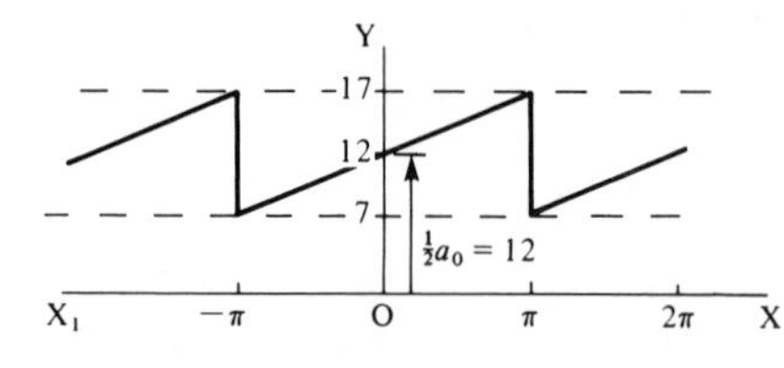

If we now raise the waveform to its original position through 12 units, i.e. $\frac{1}{2}a_0 = 12$, the series finally becomes

$$f(x) = 12 + \frac{10}{\pi}\left\{\sin x - \frac{1}{2}\sin 2x + \frac{1}{3}\sin 3x - \frac{1}{4}\sin 4x + \ldots\right\}$$

Note: In the analysis of electrical oscillations, the constant term $\frac{1}{2}a_0$ represents the d.c. component of the current or voltage.

Exercise 15

For each of the waveforms shown, determine the Fourier series to represent the function. In each case $f(x) = f(x + 2\pi)$.

1.

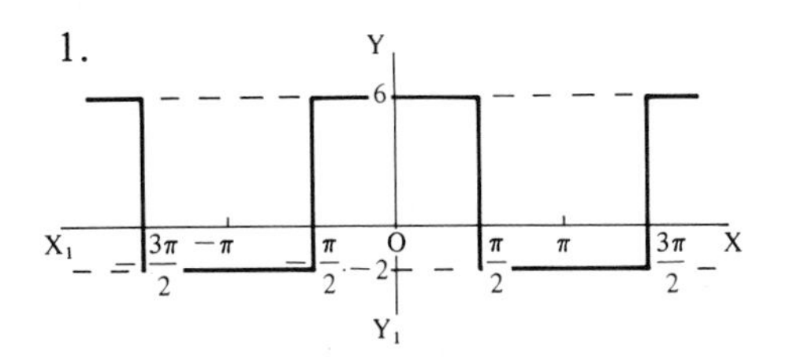

2.

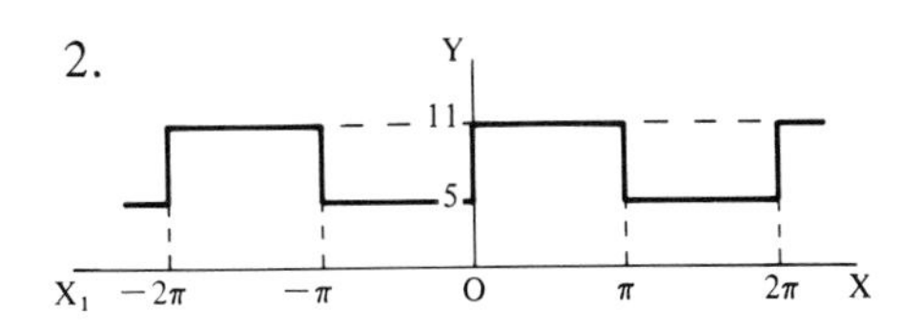

3.

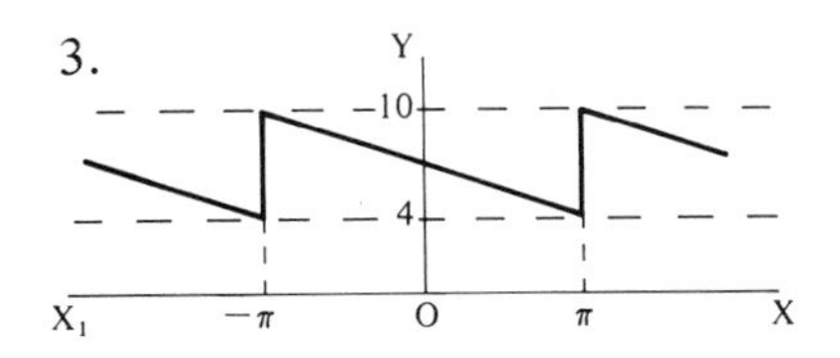

4.

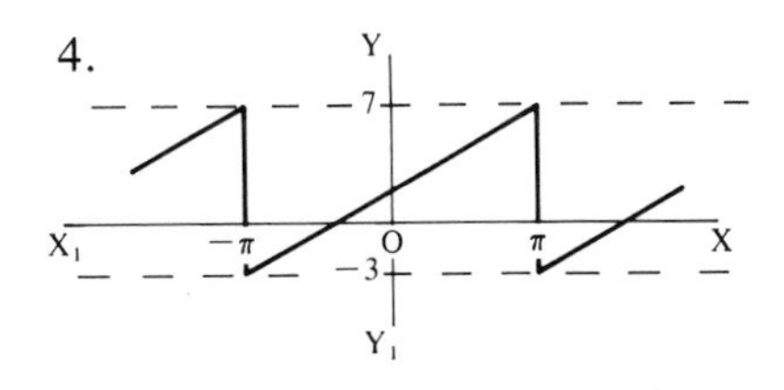

5.

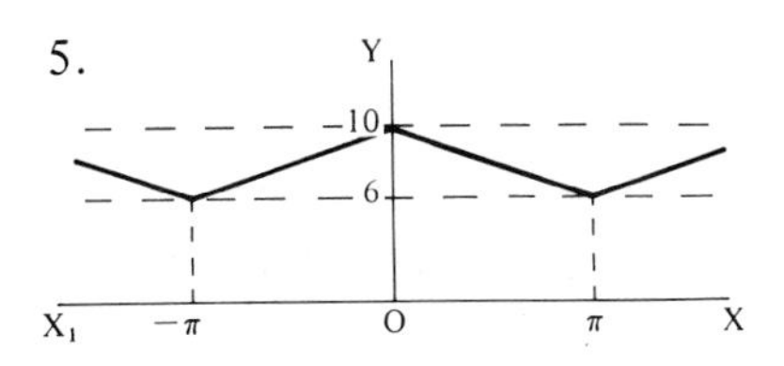

6.

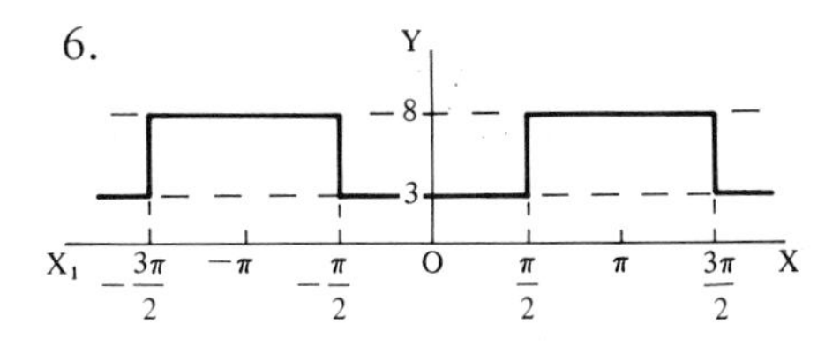

7.

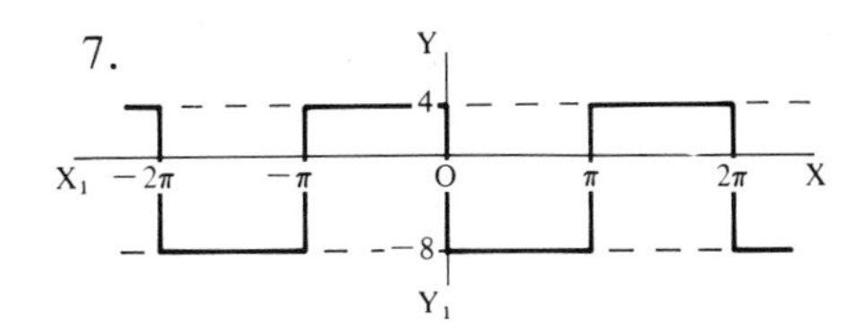

8.

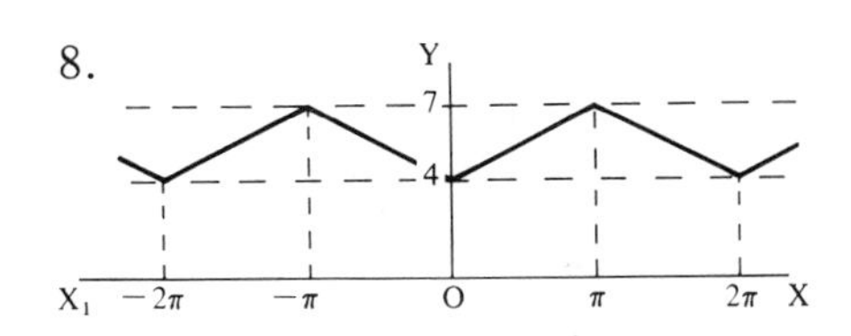

9.

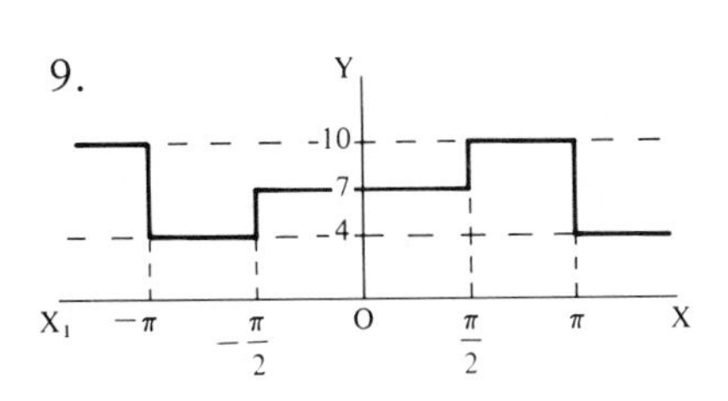

10.

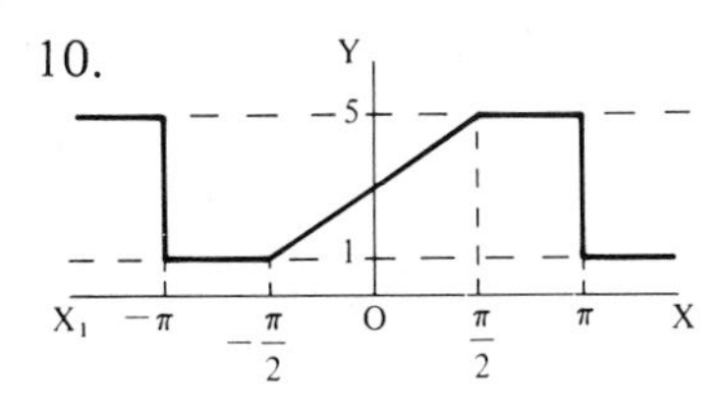

4.4 REVISION SUMMARY

1. Half-range series – functions defined over half a period

A function $f(x)$ of period 2π, defined between $x = 0$ and $x = \pi$, can be represented as a *half-range cosine series*

$$f(x) = \tfrac{1}{2}a_0 + \sum_{n=1}^{\infty} a_n \cos nx$$

where
$$a_0 = \frac{2}{\pi}\int_0^{\pi} f(x)\,dx$$

$$a_n = \frac{2}{\pi}\int_0^{\pi} f(x)\cos nx\,dx$$

or as a *half-range sine series*

$$f(x) = \sum_{n=1}^{\infty} b_n \sin nx$$

where
$$b_n = \frac{2}{\pi}\int_0^{\pi} f(x)\sin nx\,dx$$

In each case, the resulting series is valid for $0 < x < \pi$.

2. Odd and even functions

Even functions:	$f(x) = f(-x)$	cosine terms only
Odd functions:	$f(x) = -f(-x)$	sine terms only

3. Significance of the constant term $\frac{1}{2}a_0$

This signifies a vertical shift of the waveform of $f(x)$. It does not depend on the independent variable (x) and, in electrical applications, therefore, it represents a d.c. component of the voltage or current expressed as the series.

4. Odd or even harmonics only

If $f(x) = f(x + \pi)$, the Fourier series contains *even* harmonics only.

If $f(x) = -f(x + \pi)$, the Fourier series contains *odd* harmonics only.

Chapter 5

FUNCTIONS WITH PERIODS OTHER THAN 2π

5.1 FUNCTIONS WITH PERIOD $2L$

5.1.1 Change of units

So far, we have considered functions $f(x)$ with period 2π and, in certain cases π, i.e. $f(x) = f(x + 2\pi)$ and $f(x) = f(x + \pi)$. In practice, we often need to find a Fourier series for a function which is defined over an interval other than 2π, e.g. from $-L$ to L, or from 0 to $2L$, i.e. the period is $2L$.

If $y = f(x)$ is defined in the range $-L$ to L and has period $2L$, we can convert this into an interval of 2π by changing the units of the indefinite variable by the substitution $x = \dfrac{Lu}{\pi}$, i.e. $u = \dfrac{\pi x}{L}$. Then

$$\text{when} \quad x = -L, \qquad u = -\pi$$

$$\text{when} \quad x = L, \qquad u = \pi$$

The original function $\quad y = f(x) \qquad -L < x < L$

now becomes $\quad y = f\left(\dfrac{Lu}{\pi}\right) = F(u) \qquad -\pi < u < \pi$

$F(u)$ can now be expanded as a Fourier series in the usual way.

$$F(u) = \tfrac{1}{2}a_0 + \sum_{n=1}^{\infty} \{a_n \cos nu + b_n \sin nu\}$$

where

$$a_0 = \frac{1}{\pi}\int_{-\pi}^{\pi} F(u)\,\mathrm{d}u$$

$$a_n = \frac{1}{\pi}\int_{-\pi}^{\pi} F(u)\cos nu\,\mathrm{d}u$$

$$b_n = \frac{1}{\pi}\int_{-\pi}^{\pi} F(u)\sin nu\,\mathrm{d}u$$

Then, having obtained the expressions for the coefficients in terms of u, we can revert to the original variable by putting $u = \dfrac{\pi x}{L}$.

$$\therefore\ f(x) = \tfrac{1}{2}a_0 + \sum_{n=1}^{\infty}\left\{a_n \cos\frac{n\pi}{L}x + b_n \sin\frac{n\pi}{L}x\right\}$$

Example 1

To find the Fourier series for the function

$$f(x) = 2x \qquad -5 < x < 5$$

$$f(x) = f(x+10)$$

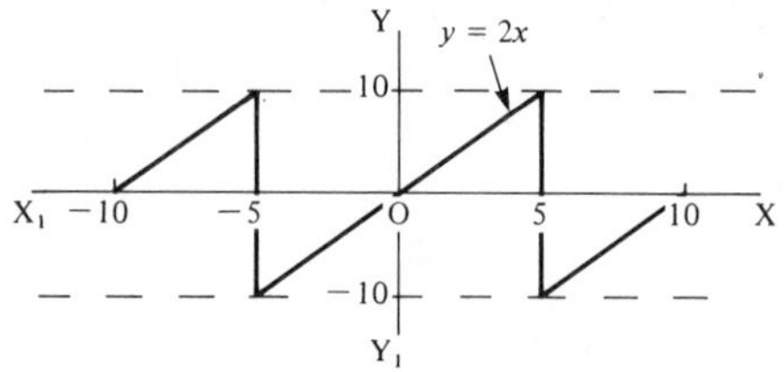

$f(x)$ is symmetrical about the origin, i.e. it is an odd function. Therefore, the series contains only sine terms.

$$f(x) = \tfrac{1}{2}a_0 + \sum_{n=1}^{\infty}\{a_n \cos nx + b_n \sin nx\}$$

In this case, $a_0 = 0$ and $a_n = 0$ $\therefore\ f(x) = \sum_{n=1}^{\infty} b_n \sin nx.$

Period $= 2L = 10$ $\therefore\ L = 5.$ Put $x = \dfrac{5u}{\pi}$ i.e. $u = \dfrac{\pi x}{5}.$

When $x = -5,\ u = -\pi$

$x = 5,\ u = \pi$

$$f(x) = f\left(\frac{5u}{\pi}\right) = F(u)$$

Then $$F(u) = \sum_{n=1}^{\infty} b_n \sin nu \qquad -\pi < u < \pi$$

$$b_n = \frac{1}{\pi}\int_{-\pi}^{\pi} F(u)\ \sin nu\ du = \frac{2}{\pi}\int_0^{\pi} F(u) \sin nu\ du$$

$$= \frac{2}{\pi}\int_0^{\pi} \frac{10u}{\pi} \sin nu\ du = \frac{20}{\pi^2}\int_0^{\pi} u \sin nu\ du$$

$$\therefore\ \frac{\pi^2 b_n}{20} = \left[u\left(\frac{-\cos nu}{n}\right)\right]_0^{\pi} + \frac{1}{n}\int_0^{\pi} \cos nu\ du$$

$$= \frac{1}{n}\Big[-u \cos nu\Big]_0^{\pi} + \frac{1}{n^2}\Big[\sin nu\Big]_0^{\pi}$$

$$= -\frac{1}{n}\pi \cos n\pi$$

$$\therefore\ b_n = -\frac{20}{n\pi}\cos n\pi \qquad \cos n\pi = -1 \quad (n \text{ odd})$$

$$= 1 \quad (n \text{ even})$$

$$\therefore\ F(u) = -\frac{20}{\pi}\left\{-\sin u + \frac{1}{2}\sin 2u - \frac{1}{3}\sin 3u + \ldots\right\}$$

Now put $u = \dfrac{\pi x}{5}$

$$\therefore\ f(x) = \frac{20}{\pi}\left\{\sin\frac{\pi x}{5} - \frac{1}{2}\sin\frac{2\pi x}{5} + \frac{1}{3}\sin\frac{3\pi x}{5} - \dots\right\}$$

Exercise 16

In each of the following cases, state the substitution necessary in the independent variable to convert the periodic interval to 2π.

1. $f(x) = 2x + 1 \qquad -4 < x < 4$
2. $f(x) = x^2 \qquad -1 < x < 1$
3. $f(x) = 1 - x^2 \qquad -10 < x < 10$
4. $f(x) = \sin\dfrac{x}{2} \qquad -2 < x < 2$
5. $f(x) = \dfrac{e^{x}}{10} \qquad -3 < x < 3$

Example 2

Determine the Fourier series for the function defined by

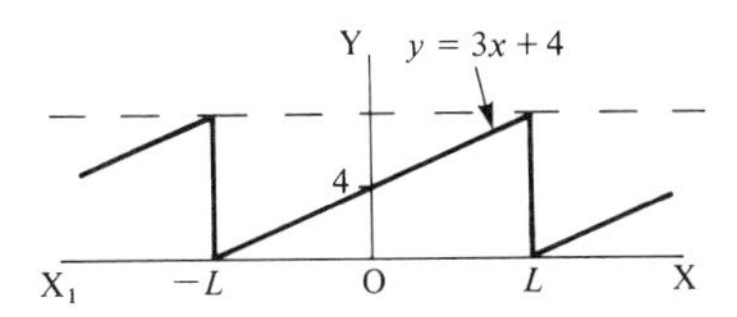

$$f(x) = 3x + 4 \qquad -L < x < L$$
$$f(x) = f(x + 2L)$$

$$f(x) = \tfrac{1}{2}a_0 + \sum_{n=1}^{\infty}\{a_n \cos nx + b_n \sin nx\}$$

If we lower the waveform through 4 units, we get

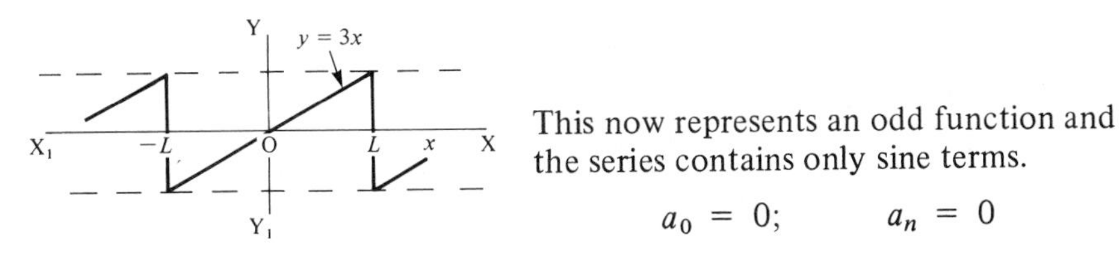

This now represents an odd function and the series contains only sine terms.

$$a_0 = 0; \qquad a_n = 0$$

Now put $x = \dfrac{Lu}{\pi}$, i.e. $u = \dfrac{\pi x}{L}$.

When $x = -L, \qquad u = -\pi$

$x = L, \qquad u = \pi$

$$f(x) = f\left(\frac{Lu}{\pi}\right) = F(u)$$

$$a_0 = 0; \qquad a_n = 0$$

$$F(u) = \sum_{n=1}^{\infty} b_n \sin nu$$

where $$b_n = \frac{2}{\pi}\int_0^{\pi} F(u)\ \sin nu\, du$$

$$\frac{\pi b_n}{2} = \left[\frac{3Lu}{\pi}\left(\frac{-\cos nu}{n}\right)\right]_0^{\pi} + \frac{3L}{\pi n}\int_0^{\pi} \cos nu\, du$$

$$= \left\{-\frac{3L\cos n\pi}{n} - 0\right\} + \frac{3L}{\pi n}\left[\frac{\sin nu}{n}\right]_0^{\pi}$$

$$= -\frac{3L}{n}\cos n\pi + 0$$

$$\therefore\ b_n = -\frac{6L}{\pi n}\cos n\pi \qquad \cos n\pi = -1 \quad (n \text{ odd})$$
$$= 1 \quad (n \text{ even})$$

$$\therefore\ F(u) = \frac{6L}{\pi}\left\{\sin u - \frac{1}{2}\sin 2u + \frac{1}{3}\sin 3u - \ldots\right\}$$

But $u = \dfrac{\pi x}{L}$

$$\therefore\ f(x) = \frac{6L}{\pi}\left\{\sin\frac{\pi x}{L} - \frac{1}{2}\sin\frac{2\pi x}{L} + \frac{1}{3}\sin\frac{3\pi x}{L} - \ldots\right\}$$

Therefore, raising the waveform to its original level, the constant term

$$\tfrac{1}{2}a_0 = 4$$

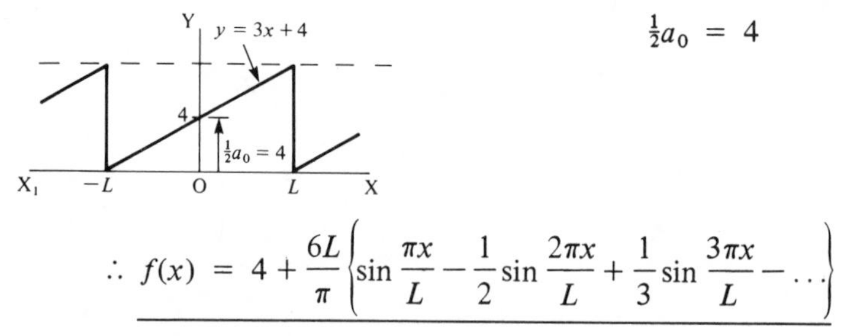

$$\therefore\ f(x) = 4 + \frac{6L}{\pi}\left\{\sin\frac{\pi x}{L} - \frac{1}{2}\sin\frac{2\pi x}{L} + \frac{1}{3}\sin\frac{3\pi x}{L} - \ldots\right\}$$

5.1.2 Alternative method

Instead of making the substitutions from first principles, in practice we can replace u by $\dfrac{\pi x}{L}$ and work directly in terms of x.

$$f(x) = F(u) = \tfrac{1}{2}a_0 + \sum_{n=1}^{\infty}\{a_n \cos nu + b_n \sin nu\}$$

then becomes $$f(x) = \tfrac{1}{2}a_0 + \sum_{n=1}^{\infty}\left\{a_n \cos\frac{n\pi x}{L} + b_n \sin\frac{n\pi x}{L}\right\}$$

where $$a_0 = \frac{1}{L}\int_{-L}^{L} f(x)\,\mathrm{d}x$$

$$a_n = \frac{1}{L}\int_{-L}^{L} f(x)\cos\frac{n\pi x}{L}\,\mathrm{d}x$$

$$b_n = \frac{1}{L}\int_{-L}^{L} f(x)\sin\frac{n\pi x}{L}\,\mathrm{d}x$$

Similarly,

(a) for a *half-range cosine series* for $y = f(x),\ 0 < x < L$

$$f(x) = \tfrac{1}{2}a_0 + \sum_{n=1}^{\infty} a_n \cos\frac{n\pi x}{L}$$

where $$a_0 = \frac{2}{L}\int_0^L f(x)\,\mathrm{d}x$$

and $$a_n = \frac{2}{L}\int_0^L f(x)\cos\frac{n\pi x}{L}\,\mathrm{d}x$$

(b) for a *half-range sine series* for $y = f(x),\ 0 < x < L$

$$f(x) = \sum_{n=1}^{\infty} b_n \sin\frac{n\pi x}{L}$$

where $$b_n = \frac{2}{L}\int_0^L f(x)\sin\frac{n\pi x}{L}\,\mathrm{d}x$$

Example 1

Determine the Fourier series for the function shown.

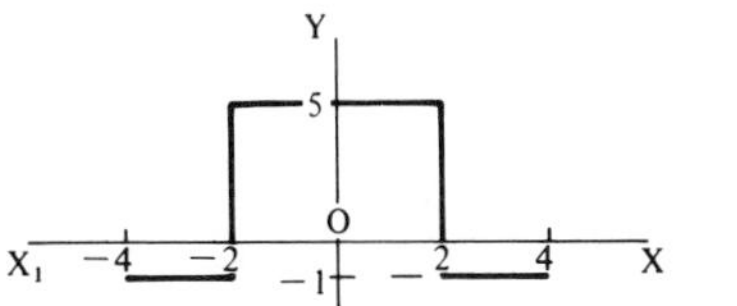

$$\begin{aligned} f(x) &= -1 & -4 < x < -2 \\ f(x) &= 5 & -2 < x < 2 \\ f(x) &= -1 & 2 < x < 4 \\ f(x) &= f(x+8) & \end{aligned}$$

First, consider the waveform lowered by 2 units.

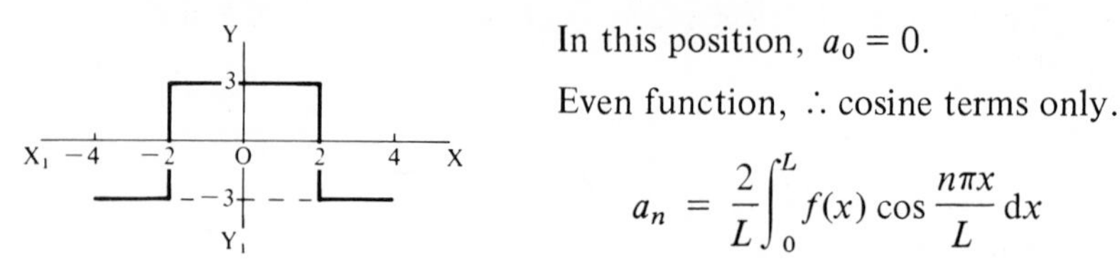

In this position, $a_0 = 0$.

Even function, $\therefore$ cosine terms only.

$$a_n = \frac{2}{L}\int_0^L f(x)\cos\frac{n\pi x}{L}\,dx$$

Period $= 2L = 8 \quad \therefore \; L = 4.$

$$a_n = \frac{1}{2}\int_0^4 f(x)\cos\frac{n\pi x}{4}\,dx$$

$$\therefore \; 2a_n = \int_0^2 3\cos\frac{n\pi x}{4}\,dx + \int_2^4 (-3)\cos\frac{n\pi x}{4}\,dx$$

$$\therefore \; \tfrac{2}{3}a_n = \frac{4}{n\pi}\left[\sin\frac{n\pi x}{4}\right]_0^2 - \frac{4}{n\pi}\left[\sin\frac{n\pi x}{4}\right]_2^4$$

$$= \frac{4}{n\pi}\left\{\sin\frac{n\pi}{2} - 0 - \sin n\pi + \sin\frac{n\pi}{2}\right\} \qquad n = 1, 2, 3, \ldots$$

$$\therefore \; a_n = \frac{12}{n\pi}\sin\frac{n\pi}{2}$$

n	1	2	3	4	5	6	7
$\sin\frac{n\pi}{2}$	1	0	−1	0	1	0	−1

$$\therefore \; f(x) = \frac{12}{\pi}\left\{\cos\frac{\pi x}{4} - \frac{1}{3}\cos\frac{3\pi x}{4} + \frac{1}{5}\cos\frac{5\pi x}{4} - \ldots\right\}$$

Restoring the waveform to its original position, i.e. raising it by 2 units,

$$\tfrac{1}{2}a_0 = 2$$

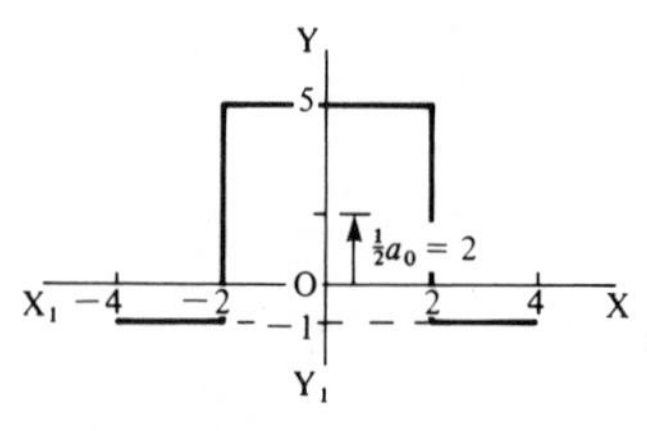

The required series is therefore

$$f(x) = 2 + \frac{12}{\pi}\left\{\cos\frac{\pi x}{4} - \frac{1}{3}\cos\frac{3\pi x}{4} + \frac{1}{5}\cos\frac{5\pi x}{4} - \ldots\right\}$$

Example 2

A periodic function $f(x)$ has a period of 6 units and is defined over the interval $0<x<3$ by

$$f(x) = 5x \qquad 0<x<2$$
$$f(x) = 10 \qquad 2<x<3$$

Determine a half-range cosine series to represent the function.

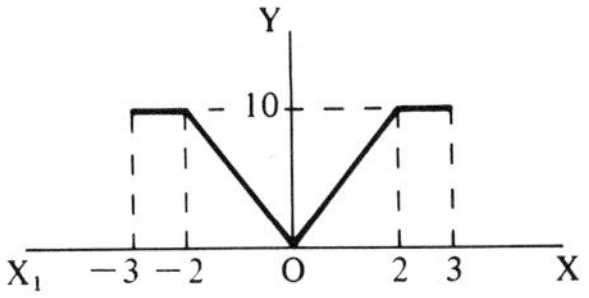

Period $= 2L = 6 \quad \therefore\ L = 3.$

$$f(x) = \tfrac{1}{2}a_0 + \sum_{n=1}^{\infty}\left\{a_n \cos\frac{n\pi x}{L} + b_n \sin\frac{n\pi x}{L}\right\}$$
$$= \tfrac{1}{2}a_0 + \sum_{n=1}^{\infty}\left\{a_n \cos\frac{n\pi x}{3} + b_n \sin\frac{n\pi x}{3}\right\}$$

Even function, $\therefore$ no sine terms $\therefore\ b_n = 0.$

$$a_0 = \frac{2}{L}\int_0^L f(x)\,\mathrm{d}x = \frac{2}{3}\left\{\int_0^2 5x\,\mathrm{d}x + \int_2^3 10\,\mathrm{d}x\right\}$$
$$= \frac{2}{3}\left\{\left[\frac{5x^2}{2}\right]_0^2 + \left[10x\right]_2^3\right\}$$
$$= \frac{2}{3}\{10-0+30-20\} = \frac{40}{3} \qquad \therefore\ \tfrac{1}{2}a_0 = \frac{20}{3}$$

$$a_n = \frac{2}{L}\int_0^L f(x)\cos\frac{n\pi x}{L}\,\mathrm{d}x$$
$$= \frac{2}{3}\left\{\int_0^2 5x\cos\frac{n\pi x}{3}\,\mathrm{d}x + \int_2^3 10\cos\frac{n\pi x}{3}\,\mathrm{d}x\right\}$$
$$= \frac{2}{3}\left\{\left[\frac{15x}{n\pi}\sin\frac{n\pi x}{3}\right]_0^2 - \frac{15}{n\pi}\int_0^2 \sin\frac{n\pi x}{3}\,\mathrm{d}x + \frac{30}{n\pi}\left[\sin\frac{n\pi x}{3}\right]_2^3\right\}$$
$$= \frac{2}{3}\left\{\frac{30}{n\pi}\sin\frac{2n\pi}{3} - 0 - \frac{45}{n^2\pi^2}\left[-\cos\frac{n\pi x}{3}\right]_0^2 + \frac{30}{n\pi}\left[\sin n\pi - \sin\frac{2n\pi}{3}\right]\right\}$$
$$= -\frac{30}{n^2\pi^2}\left\{1 - \cos\frac{2n\pi}{3}\right\}$$

n	1	2	3	4	5	6
$\cos\frac{2n\pi}{3}$	$-\frac{1}{2}$	$-\frac{1}{2}$	1	$-\frac{1}{2}$	$-\frac{1}{2}$	1

$$\therefore\ f(x) = \frac{20}{3} + \frac{30}{\pi^2}\left\{\frac{1}{2}\cos\frac{\pi x}{3} + \frac{1}{8}\cos\frac{2\pi x}{3} - \frac{1}{9}\cos\frac{3\pi x}{3} + \frac{1}{32}\cos\frac{4\pi x}{3}\ldots\right\}$$

Exercise 17

Determine Fourier series for the following functions.

1.

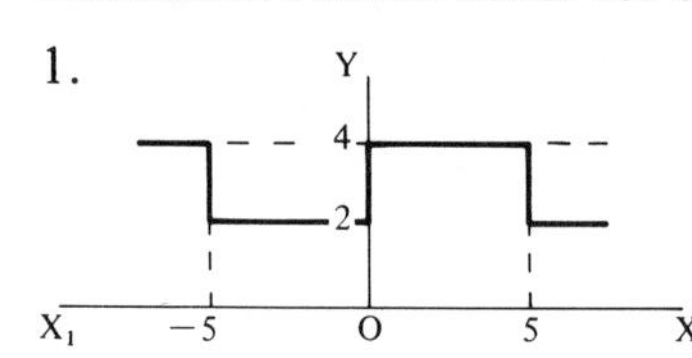

$f(x) = 2 \qquad -5 < x < 0$

$f(x) = 4 \qquad 0 < x < 5$

$f(x) = f(x + 10)$

2.

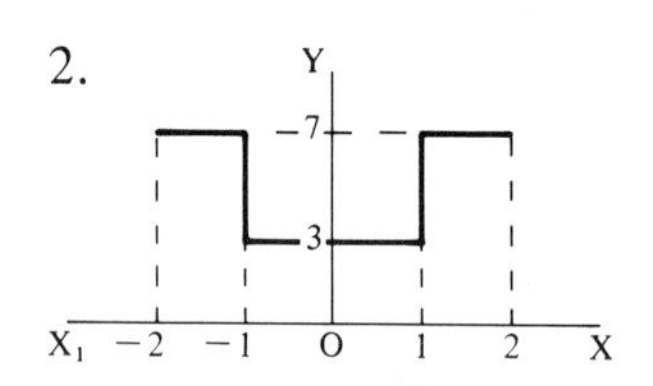

$f(x) = 7 \qquad -2 < x < -1$

$f(x) = 3 \qquad -1 < x < 1$

$f(x) = 7 \qquad 1 < x < 2$

$f(x) = f(x + 4)$

3.

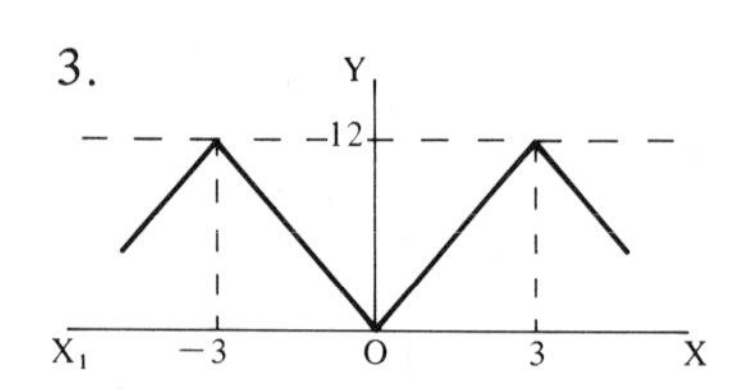

$f(x) = -4x \qquad -3 < x < 0$

$f(x) = 4x \qquad 0 < x < 3$

$f(x) = f(x + 6)$

4.

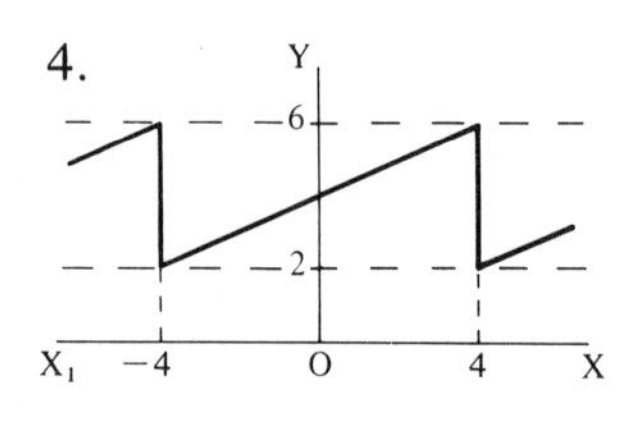

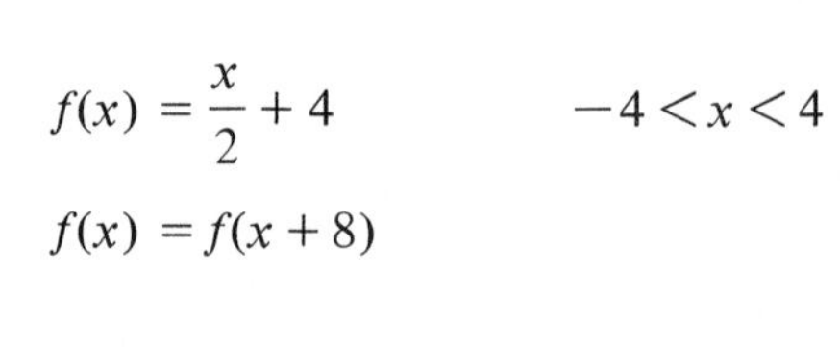

$f(x) = \dfrac{x}{2} + 4 \qquad -4 < x < 4$

$f(x) = f(x + 8)$

5.

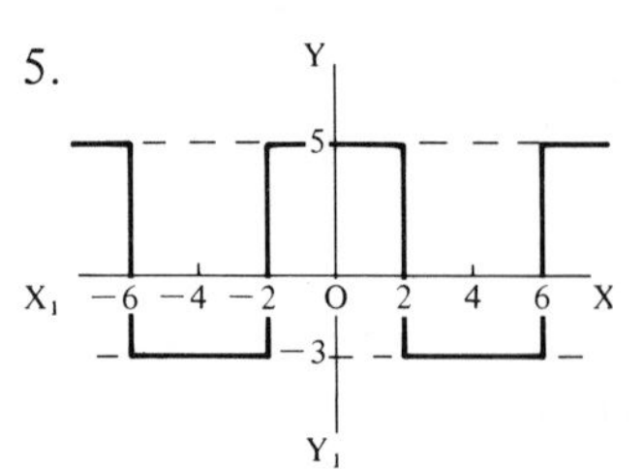

$f(x) = -3 \qquad -4 < x < -2$

$f(x) = 5 \qquad -2 < x < 2$

$f(x) = -3 \qquad 2 < x < 4$

$f(x) = f(x + 8)$

6.

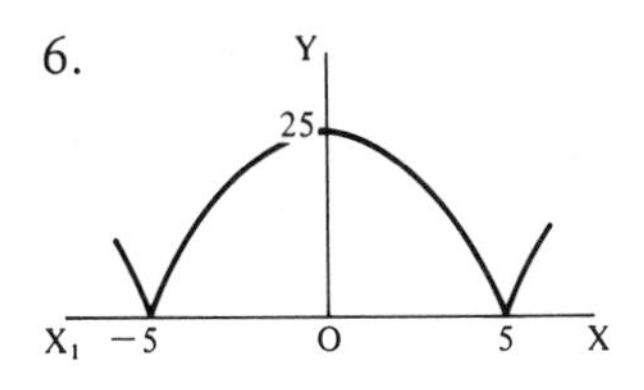

$$f(x) = 25 - x^2 \qquad -5 < x < 5$$
$$f(x) = f(x+10)$$

5.2 FUNCTIONS WITH PERIOD T

In many cases, periodic functions arise in branches of technology as physical oscillations, mechanical vibrations, or electrical or electronic oscillations, where time t is the independent variable and where the periodic interval is normally denoted by T,

i.e. $$f(t) = f(t+T)$$

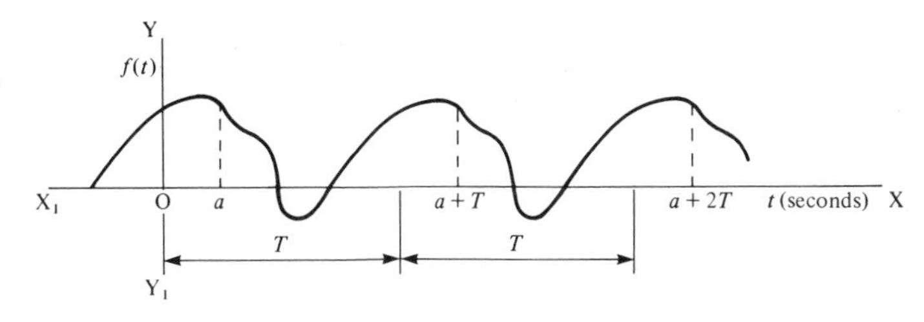

Since each cycle is completed in T seconds, the *frequency*, f hertz (or oscillations per second) of the periodic function is therefore given by $f = \frac{1}{T}$.

If the angular velocity, ω radians per second, is defined by $\omega = 2\pi f$, then $\omega = \frac{2\pi}{T}$ and $T = \frac{2\pi}{\omega}$.

The angle, x radians, at any time t seconds is therefore $x = \omega t$ and the Fourier series to represent the function can be expressed in the form

$$f(t) = \tfrac{1}{2}a_0 + \sum_{n=1}^{\infty} \{a_n \cos n\omega t + b_n \sin n\omega t\}$$

or $$f(t) = \tfrac{1}{2}A_0 + \sum_{n=1}^{\infty} B_n \sin(n\omega t + \phi_n) \qquad n = 1, 2, 3, \ldots$$

where $A_0 = a_0$; $\quad B_n \sin\phi_n = a_n$; $\quad B_n \cos\phi_n = b_n$.

Therefore $$B_n = \sqrt{a_n^2 + b_n^2}$$

and $$\phi_n = \arctan\left(\frac{a_n}{b_n}\right)$$

$B_1 \sin(\omega t + \phi_1)$ is the *first harmonic* or *fundamental* (lowest frequency)

$B_2 \sin(2\omega t + \phi_2)$ is the *second harmonic* (frequency twice that of the fundamental)

$B_n \sin(n\omega t + \phi_n)$ is the nth harmonic (frequency n times that of the fundamental)

For the series $f(t) = \frac{1}{2}A_0 + \sum_{n=1}^{\infty} B_n \sin(n\omega t + \phi_n)$ to converge, the value of B_n must eventually decrease with higher order harmonics, i.e. $B_n \to 0$ as $n \to \infty$.

5.2.1 Fourier coefficients

$$f(t) = \tfrac{1}{2}a_0 + \sum_{n=1}^{\infty} \{a_n \cos n\omega t + b_n \sin n\omega t\}$$

$$a_0 = \frac{2}{T}\int_0^T f(t)\,\mathrm{d}t = \frac{\omega}{\pi}\int_0^{2\pi/\omega} f(t)\,\mathrm{d}t$$

$$a_n = \frac{2}{T}\int_0^T f(t)\cos n\omega t\,\mathrm{d}t = \frac{\omega}{\pi}\int_0^{2\pi/\omega} f(t)\cos n\omega t\,\mathrm{d}t$$

$$b_n = \frac{2}{T}\int_0^T f(t)\sin n\omega t\,\mathrm{d}t = \frac{\omega}{\pi}\int_0^{2\pi/\omega} f(t)\sin n\omega t\,\mathrm{d}t$$

As with previous work, the integration can be carried out over any complete period, i.e. 0 to T; $-\frac{T}{2}$ to $\frac{T}{2}$; $-\frac{\pi}{\omega}$ to $\frac{\pi}{\omega}$; 0 to $\frac{2\pi}{\omega}$; etc.

5.2.2 Half-range series

(a) *Even function* Half-range cosine series

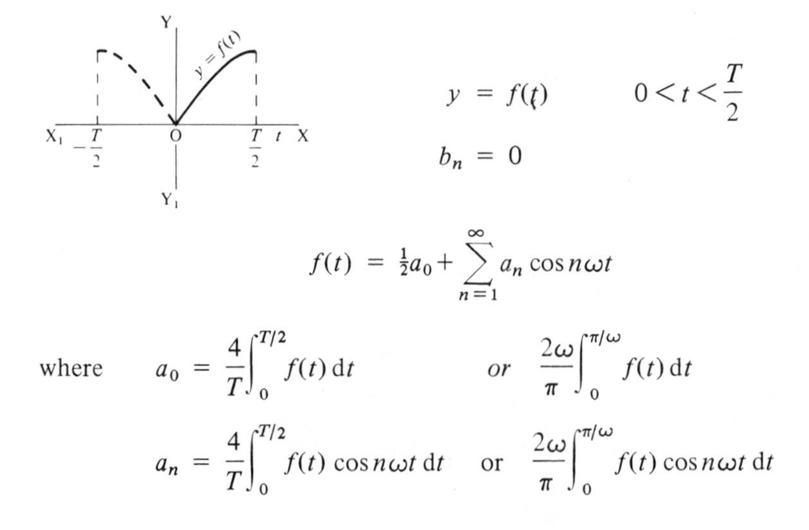

$$y = f(t) \qquad 0 < t < \frac{T}{2}$$

$$b_n = 0$$

$$f(t) = \tfrac{1}{2}a_0 + \sum_{n=1}^{\infty} a_n \cos n\omega t$$

where $a_0 = \frac{4}{T}\int_0^{T/2} f(t)\,\mathrm{d}t$ *or* $\frac{2\omega}{\pi}\int_0^{\pi/\omega} f(t)\,\mathrm{d}t$

$$a_n = \frac{4}{T}\int_0^{T/2} f(t)\cos n\omega t\,\mathrm{d}t \quad \text{or} \quad \frac{2\omega}{\pi}\int_0^{\pi/\omega} f(t)\cos n\omega t\,\mathrm{d}t$$

(b) *Odd function* Half-range sine series

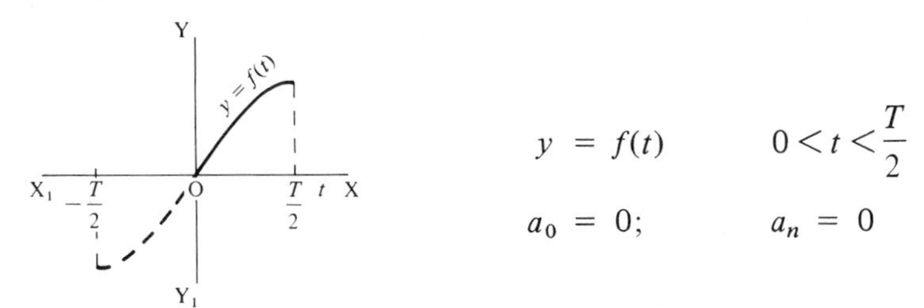

$$y = f(t) \qquad 0 < t < \frac{T}{2}$$

$$a_0 = 0; \qquad a_n = 0$$

$$f(t) = \sum_{n=1}^{\infty} b_n \sin n\omega t$$

where $$b_n = \frac{4}{T}\int_0^{T/2} f(t)\ \sin n\omega t\, dt \quad \text{or} \quad \frac{2\omega}{\pi}\int_0^{\pi/\omega} f(t)\ \sin n\omega t\, dt$$

Example 1

Determine the Fourier series to represent the function shown.

$$f(t) = 1 \qquad -2 < t < -1$$
$$f(t) = 0 \qquad -1 < t < 1$$
$$f(t) = -1 \qquad 1 < t < 2$$
$$f(t) = f(t+4)$$

Period = 4 $\therefore\ T = 4$.

Symmetrical about the origin; therefore sine terms only.

$$f(t) = \sum_{n=1}^{\infty} b_n \sin n\omega t \qquad \omega = 2\pi f = 2\pi\frac{1}{T} = 2\pi\frac{1}{4} = \frac{\pi}{2}$$

$$b_n = \frac{4}{T}\int_0^2 f(t) \sin n\omega t\, dt$$

$$= \frac{4}{T}\int_0^1 0 \sin n\omega t\, dt + \int_1^2 (-1) \sin n\omega t\, dt$$

$$= \frac{4}{4}\left[\frac{\cos n\omega t}{n\omega}\right]_1^2 = \frac{1}{n\omega}\{\cos 2n\omega - \cos n\omega\} \qquad \text{But } \omega = \frac{\pi}{2}$$

$$= \frac{2}{n\pi}\left\{\cos n\pi - \cos\frac{n\pi}{2}\right\} \qquad n = 1, 2, 3, \ldots$$

$$= \frac{2}{n\pi}\left\{(-1)^n - \cos\frac{n\pi}{2}\right\}$$

Putting $n = 1, 2, 3, \ldots,$ we have

n	1	2	3	4	5	6	7
b_n	$-\frac{2}{\pi}$	$\frac{2}{\pi}$	$-\frac{2}{3\pi}$	0	$-\frac{2}{5\pi}$	$\frac{2}{3\pi}$	$-\frac{2}{7\pi}$

$$\therefore\ f(t) = -\frac{2}{\pi}\left\{\sin\omega t - \sin 2\omega t + \frac{1}{3}\sin 3\omega t + \frac{1}{5}\sin 5\omega t - \frac{1}{3}\sin 6\omega t + \frac{1}{7}\sin 7\omega t + \ldots\right\}$$

where $\omega = \frac{\pi}{2}$.

Example 2

A function $f(t)$ is defined by $f(t) = -\frac{4}{5}t + 4$ in the interval $t = 0$ to $t = 5$. Form a half-range cosine series to represent the function in this interval.

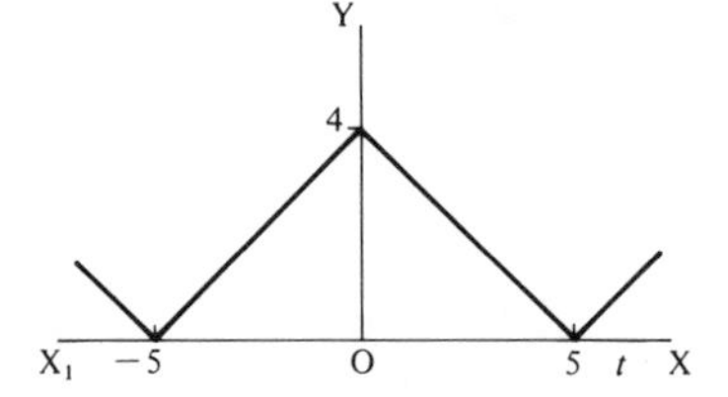

Form an even function, i.e. symmetrical about the y-axis.

Lowering the waveform through 2 units gives

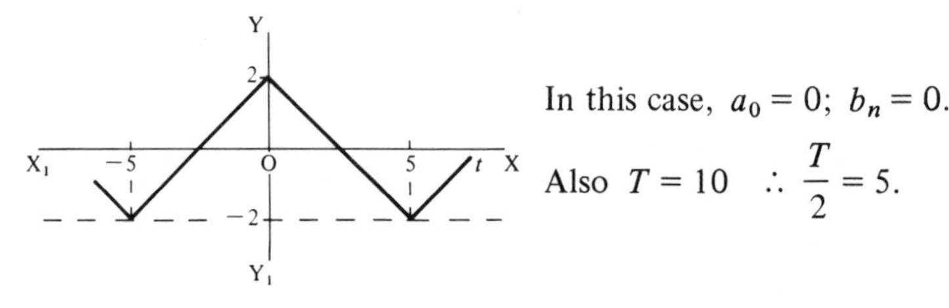

In this case, $a_0 = 0$; $b_n = 0$.

Also $T = 10 \quad \therefore \frac{T}{2} = 5.$

$$a_n = \frac{4}{T}\int_0^{T/2} f(t)\cos n\omega t\,dt = \frac{4}{10}\int_0^5 (-\tfrac{4}{5}t + 2)\cos n\omega t\,dt$$

$$= \frac{2}{5}\left\{\left[(-\tfrac{4}{5}t + 2)\frac{\sin n\omega t}{n\omega}\right]_0^5 + \frac{4}{5n\omega}\int_0^5 \sin n\omega t\,dt\right\}$$

$$= \frac{2}{5}\left\{\frac{-2\sin 5n\omega}{n\omega} + \frac{4}{5n\omega}\left[\frac{-\cos n\omega t}{n\omega}\right]_0^5\right\}$$

$$= \frac{2}{5}\left\{\frac{-2\sin 5n\omega}{n\omega} - \frac{4}{5n^2\omega^2}(\cos 5n\omega - 1)\right\}$$

$$= -\frac{4}{5}\frac{\sin 5n\omega}{n\omega} + \frac{8}{25n^2\omega^2}(1-\cos 5n\omega) \qquad \omega = \frac{2\pi}{10} = \frac{\pi}{5}$$

$$= -\frac{4}{5}\frac{\sin n\pi}{n(\pi/5)} + \frac{8}{n^2\pi^2}(1-\cos n\pi) \qquad n = 1, 2, 3, \ldots$$

$$= 0 \quad (n \text{ even}) \quad \text{or} \quad \frac{16}{n^2\pi^2} \quad (n \text{ odd})$$

$$\therefore\ f(t) = \frac{16}{\pi^2}\left\{\cos\omega t + \frac{1}{3^2}\cos 3\omega t + \frac{1}{5^2}\cos 5\omega t + \ldots\right\}$$

Finally, raising the waveform to its original position, $\frac{1}{2}a_0 = 2$, and therefore

$$f(t) = 2 + \frac{16}{\pi^2}\left\{\cos\omega t + \frac{1}{3^2}\cos 3\omega t + \frac{1}{5^2}\cos 5\omega t + \ldots\right\} \quad \text{where } \omega = \frac{\pi}{5}.$$

Exercise 18

Determine Fourier series for the following functions.

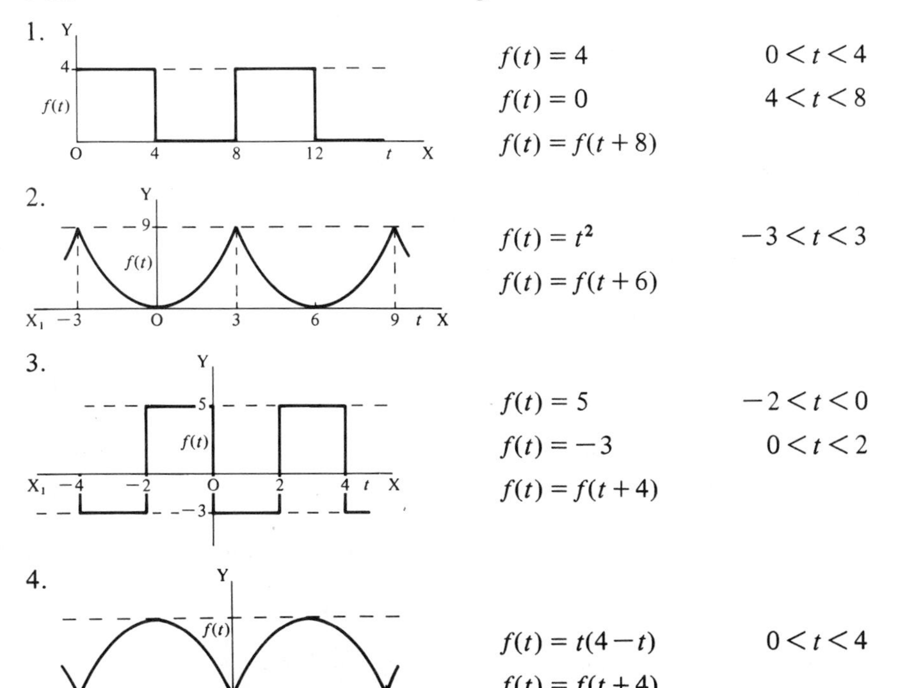

1. $f(t) = 4 \qquad 0 < t < 4$
   $f(t) = 0 \qquad 4 < t < 8$
   $f(t) = f(t+8)$

2. $f(t) = t^2 \qquad -3 < t < 3$
   $f(t) = f(t+6)$

3. $f(t) = 5 \qquad -2 < t < 0$
   $f(t) = -3 \qquad 0 < t < 2$
   $f(t) = f(t+4)$

4. $f(t) = t(4-t) \qquad 0 < t < 4$
   $f(t) = f(t+4)$

5.

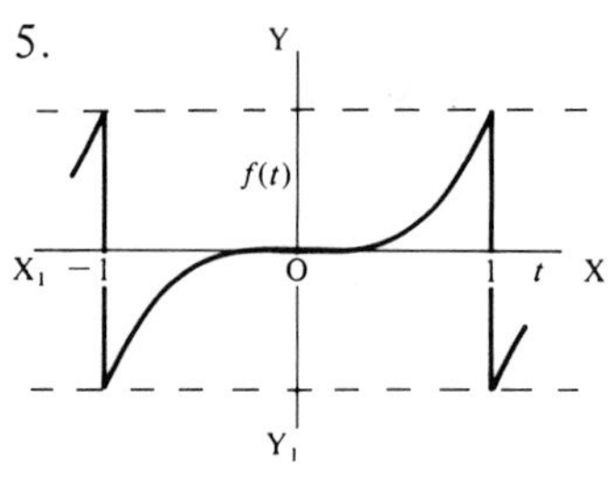

$$f(t) = t^3 \qquad -1 < t < 1$$
$$f(t) = f(t+2)$$

6.

$$f(t) = 1 \qquad -2 < t < 0$$
$$f(t) = -\frac{t}{3} + 1 \qquad 0 < t < 3$$
$$f(t) = f(t+5)$$

7.

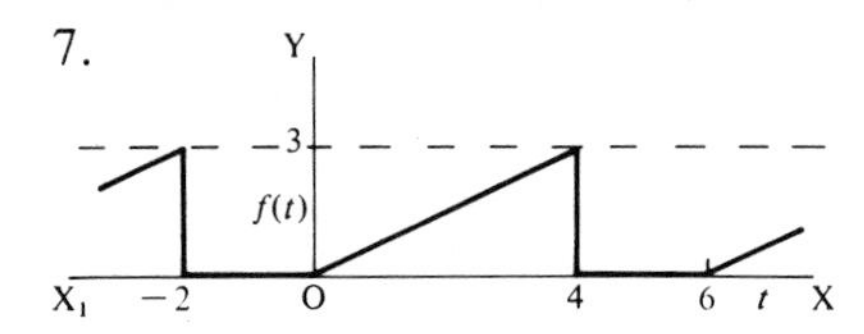

$$f(t) = 0 \qquad -2 < t < 0$$
$$f(t) = \frac{3t}{4} \qquad 0 < t < 4$$
$$f(t) = f(t+6)$$

8.

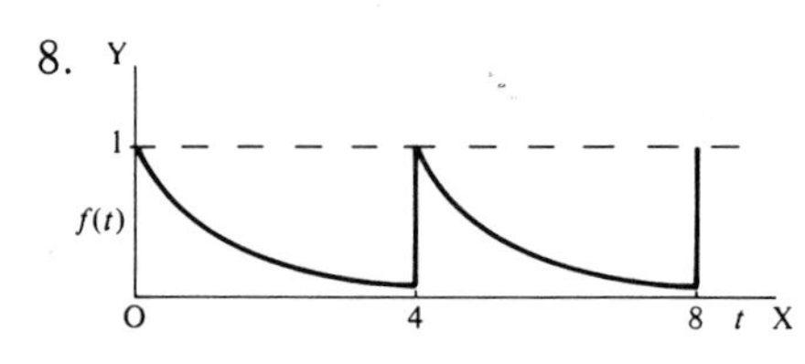

$$f(t) = e^{-t} \qquad 0 < t < 4$$
$$f(t) = f(t+4)$$

5.3 REVISION SUMMARY

Functions of periods other than 2π – change of units

1. Period 2*L*

$$f(x) = f(x+2L)$$

Put $x = \frac{Lu}{\pi}$, i.e. $u = \frac{\pi x}{L}$.

$$f(x) = \tfrac{1}{2}a_0 + \sum_{n=1}^{\infty}\{a_n \cos nu + b_n \sin nu\}$$

becomes

$$f(x) = \tfrac{1}{2}a_0 + \sum_{n=1}^{\infty}\left\{a_n \cos\frac{n\pi x}{L} + b_n \sin\frac{n\pi x}{L}\right\}$$

where
$$a_0 = \frac{1}{L}\int_{-L}^{L} f(x)\,dx$$

$$a_n = \frac{1}{L}\int_{-L}^{L} f(x)\cos\frac{n\pi x}{L}\,dx$$

$$b_n = \frac{1}{L}\int_{-L}^{L} f(x)\sin\frac{n\pi x}{L}\,dx$$

(a) *Half-range cosine series* $y = f(x),\ 0 < x < L$

$$f(x) = \tfrac{1}{2}a_0 + \sum_{n=1}^{\infty} a_n \cos\frac{n\pi x}{L}$$

where
$$a_0 = \frac{2}{L}\int_{0}^{L} f(x)\,dx$$

$$a_n = \frac{2}{L}\int_{0}^{L} f(x)\cos\frac{n\pi x}{L}\,dx$$

(b) *Half-range sine series* $y = f(x),\ 0 < x < L$

$$f(x) = \sum_{n=1}^{\infty} b_n \sin\frac{n\pi x}{L}$$

where
$$b_n = \frac{2}{L}\int_{0}^{L} f(x)\sin\frac{n\pi x}{L}\,dx$$

2. Period T

$$f(t) = f(t+T) \qquad \omega = 2\pi f = \frac{2\pi}{T} \qquad \therefore\ T = \frac{2\pi}{\omega}$$

$$f(t) = \tfrac{1}{2}a_0 + \sum_{n=1}^{\infty}\{a_n \cos n\omega t + b_n \sin n\omega t\}$$

$$a_0 = \frac{2}{T}\int_{0}^{T} f(t)\,dt \qquad = \frac{\omega}{\pi}\int_{0}^{2\pi/\omega} f(t)\,dt$$

$$a_n = \frac{2}{T}\int_{0}^{T} f(t)\cos n\omega t\,dt = \frac{\omega}{\pi}\int_{0}^{2\pi/\omega} f(t)\cos n\omega t\,dt$$

$$b_n = \frac{2}{T}\int_{0}^{T} f(t)\sin n\omega t\,dt = \frac{\omega}{\pi}\int_{0}^{2\pi/\omega} f(t)\sin n\omega t\,dt$$

(a) *Half-range cosine series* $y = f(t),\ 0 < t < \dfrac{T}{2}$

$$f(t) = \tfrac{1}{2}a_0 + \sum_{n=1}^{\infty} a_n \cos n\omega t$$

$$a_0 = \frac{4}{T}\int_0^{T/2} f(t)\,\mathrm{d}t = \frac{2\omega}{\pi}\int_0^{\pi/\omega} f(t)\,\mathrm{d}t$$

$$a_n = \frac{4}{T}\int_0^{T/2} f(t)\cos n\omega t\,\mathrm{d}t = \frac{2\omega}{\pi}\int_0^{\pi/\omega} f(t)\cos n\omega t\,\mathrm{d}t$$

(b) *Half-range sine series* $y = f(t),\ 0 < t < \dfrac{T}{2}$

$$f(t) = \sum_{n=1}^{\infty} b_n \sin n\omega t$$

$$b_n = \frac{4}{T}\int_0^{T/2} f(t)\sin n\omega t\,\mathrm{d}t = \frac{2\omega}{\pi}\int_0^{\pi/\omega} f(t)\sin n\omega t\,\mathrm{d}t$$

Chapter 6

NUMERICAL HARMONIC ANALYSIS

6.1 APPROXIMATE INTEGRATION

In experimental work, an output function is often obtained in the form of a set of readings or as a graph, rather than as an algebraic expression. Direct integration to determine the coefficients of the sine and cosine terms of the relevant Fourier series is not then possible and approximate methods must be used.

As before, the required series is denoted by

$$f(x) = \tfrac{1}{2}a_0 + \sum_{n=1}^{\infty} \{a_n \cos nx + b_n \sin nx\}$$

where

$$a_0 = \frac{1}{\pi}\int_0^{2\pi} f(x)\,\mathrm{d}x \qquad = 2 \times \text{mean value of } f(x) \text{ over a period}$$

$$a_n = \frac{1}{\pi}\int_0^{2\pi} f(x)\cos nx\,\mathrm{d}x = 2 \times \text{mean value of } f(x)\cos nx \text{ over a period}$$

$$b_n = \frac{1}{\pi}\int_0^{2\pi} f(x)\sin nx\,\mathrm{d}x = 2 \times \text{mean value of } f(x)\sin nx \text{ over a period}$$

Approximate integration, which is equivalent to finding an area, can be performed by the application of the *trapezoidal rule.*

6.1.1 Trapezoidal rule

Consider one cycle of a periodic function of period 2π, as shown.

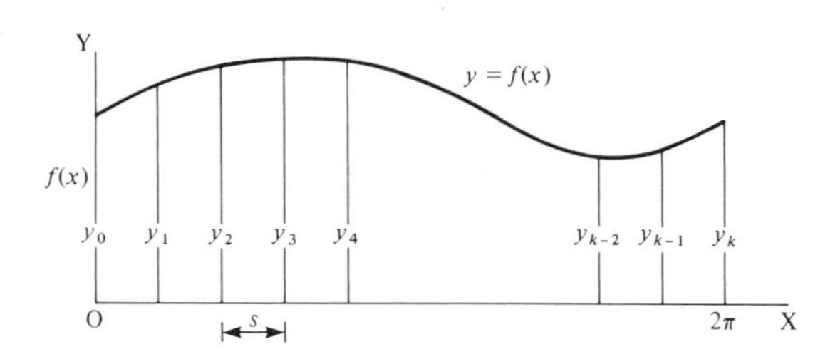

In applying the *trapezoidal rule*, the figure is divided into a convenient number (k) of strips each of width s, i.e. $\frac{2\pi}{k}$. The ordinates are then denoted by

$$y_0, y_1, y_2, \ldots, y_{k-1}, y_k$$

as indicated.

Regarding each strip as an approximate trapezium, the total area of the figure from $x = 0$ to $x = 2\pi$ is given by

$$A \approx \tfrac{1}{2}(y_0+y_1)s + \tfrac{1}{2}(y_1+y_2)s + \ldots + \tfrac{1}{2}(y_{k-2}+y_{k-1})s + \tfrac{1}{2}(y_{k-1}+y_k)s$$

$$\approx \frac{s}{2}\{(y_0+y_1)+(y_1+y_2)+\ldots+(y_{k-2}+y_{k-1})+(y_{k-1}+y_k)\}$$

$$\approx \frac{s}{2}\{y_0+2y_1+2y_2+\ldots+2y_{k-1}+y_k\}$$

$$\approx s\{\tfrac{1}{2}(y_0+y_k)+y_1+y_2+\ldots+y_{k-1}\}$$

Since $f(x)$ is periodic $y_k = y_0$

$$\therefore \int_0^{2\pi} y\,dx = A \approx s\{y_0+y_1+\ldots+y_{k-1}\}$$

where s = width of each strip

k = number of strips

6.2 TWELVE-POINT ANALYSIS

In practice, we normally divide the periodic interval (2π) into 12 equal parts, i.e. $\frac{2\pi}{12} = \frac{\pi}{6} = \frac{180^\circ}{6} = 30^\circ$, and determine from the given graph or table of readings the function value at each point of division. Note that we include the first boundary ordinate (y_0), but not the final boundary ordinate (y_k) which corresponds to the first ordinate of the next cycle.

Example 1

The graph of a periodic function $f(x)$ is shown below.

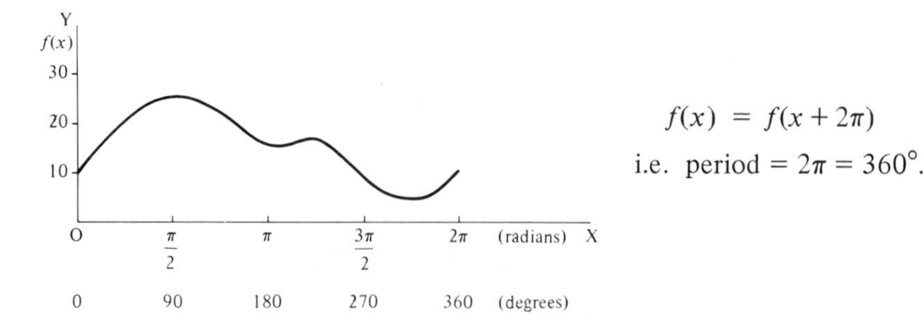

Function values at intervals of 30° are as follows:

$x°$	0	30	60	90	120	150	180	210	240	270	300	330	360
$f(x)$	10	18	24	26	25	20	16	16	15	9	5	5	10

(a) *To find a_0*

$$a_0 = 2 \times \text{mean value of } f(x) \text{ over a period}$$

$$= 2 \times \frac{1}{12} \times \text{sum of 12 ordinates} = \frac{1}{6} \times \sum_{r=0}^{11} y_r$$

$$= \frac{1}{6}\{y_0 + y_1 + y_2 + \ldots + y_{11}\}$$

$x°$	0	30	60	90	120	150	180	210	240	270	300	330	360
y	y_0	y_1	y_2	y_3	y_4	y_5	y_6	y_7	y_8	y_9	y_{10}	y_{11}	–
$f(x)$	10	18	24	26	25	20	16	16	15	9	5	5	–

$$\therefore a_0 = \frac{1}{6}\{10 + 18 + 24 + 26 + 25 + 20 + 16 + 16 + 15 + 9 + 5 + 5\}$$

$$= \frac{1}{6} \times 189 = 31.5 \qquad \therefore \underline{a_0 = 31.5}$$

(b) *To find a_1*

$$a_1 = 2 \times \text{mean value of } f(x) \cos x \text{ over a period}$$

$$= \frac{1}{6} \sum_{r=0}^{11} y_r \cos x_r$$

$x°$	0	30	60	90	120	150	180	210	240	270	300	330	360
y	y_0	y_1	y_2	y_3	y_4	y_5	y_6	y_7	y_8	y_9	y_{10}	y_{11}	–
$f(x)$	10	18	24	26	25	20	16	16	15	9	5	5	–
$\cos x$	1	0.866	0.5	0	−0.5	−0.866	−1	−0.866	−0.5	0	0.5	0.866	–
$y \cos x$	10	15.59	12	0	−12.5	−17.32	−16	−13.86	−7.5	0	2.5	4.33	–

$$a_1 = \frac{1}{6}\{10 + 15.59 + 12 + 0 - 12.5 - \ldots + 2.5 + 4.33\}$$

$$= \frac{1}{6}\{-22.76\} = -3.79 \qquad \therefore \underline{a_1 = -3.79}$$

(c) *To find a_2*

$$a_2 = 2 \times \text{mean value of } f(x) \cos 2x \text{ over a period}$$

$$= \frac{1}{6} \sum_{r=0}^{11} y_r \cos 2x_r$$

$x°$	0	30	60	90	120	150	180	210	240	270	300	330	360
y	y_0	y_1	y_2	y_3	y_4	y_5	y_6	y_7	y_8	y_9	y_{10}	y_{11}	–
$f(x)$	10	18	24	26	25	20	16	16	15	9	5	5	–
$\cos 2x$	1	0.5	−0.5	−1	−0.5	0.5	1	0.5	−0.5	−1	−0.5	0.5	–
$y \cos 2x$	10	9	−12	−26	−12.5	10	16	8	−7.5	−9	−2.5	2.5	–

$$a_2 = \frac{1}{6}\{10+9-12-26-12.5+\ldots-9-2.5+2.5\}$$

$$= \frac{1}{6}\{-14\} = -2.33 \qquad \therefore \underline{a_2 = -2.33}$$

(d) *To find b_1*

$$b_1 = 2 \times \text{mean value of } f(x) \sin x \text{ over a period}$$

$$= \frac{1}{6} \sum_{r=0}^{11} y_r \sin x_r$$

$x°$	0	30	60	90	120	150	180	210	240	270	300	330	360
y	y_0	y_1	y_2	y_3	y_4	y_5	y_6	y_7	y_8	y_9	y_{10}	y_{11}	–
$f(x)$	10	18	24	26	25	20	16	16	15	9	5	5	–
$\sin x$	0	0.5	0.866	1	0.866	0.5	0	−0.5	−0.866	−1	−0.866	−0.5	–
$y \sin x$	0	9	20.78	26	21.65	10	0	−8	−12.99	−9	−4.33	−2.5	–

$$b_1 = \frac{1}{6} \sum_{r=0}^{11} y_r \sin x_r = \frac{1}{6} \times 50.61 = 8.435 \qquad \therefore \underline{b_1 = 8.44}$$

(e) *To find b_2*

$$b_2 = 2 \times \text{mean value of } f(x) \sin 2x \text{ over a period}$$

$$= \frac{1}{6} \sum_{r=0}^{11} y_r \sin 2x_r$$

Compiling a table in the same manner gives the result $\underline{b_2 = 2.60.}$

$$\therefore\ f(x) = \tfrac{1}{2}(31.5) - 3.79\cos x - 2.33\cos 2x + \ldots$$
$$+ 8.44\sin x + 2.60\sin 2x + \ldots$$

$$\therefore\ f(x) = 15.8 - 3.79\cos x - 2.33\cos 2x + \ldots$$
$$+ 8.44\sin x + 2.60\sin 2x + \ldots$$

The coefficients of subsequent harmonics can be calculated in similar manner.

6.2.1 Tabular form

The setting out of the working can be greatly reduced by the use of a standard form of table. The development of the calculations for the same function as that in Example 1 is then as follows

$x°$	y	$\cos x$	$y\cos x$	$\cos 2x$	$y\cos 2x$	$\cos 3x$	$y\cos 3x$	$\cos 4x$	$y\cos 4x$
0	10	1	10	1	10	1	10	1	10
30	18	0.866	15.59	0.5	9	0	0	−0.5	−9.0
60	24	0.5	12.00	−0.5	−12	−1	−24	−0.5	−12.0
90	26	0	0	−1	−26	0	0	1.0	26.0
120	25	−0.5	−12.50	−0.5	−12.5	1	25	−0.5	−12.5
150	20	−0.866	−17.32	0.5	10	0	0	−0.5	−10.0
180	16	−1	−16.00	1	16	−1	−16	1.0	16.0
210	16	−0.866	−13.86	0.5	8	0	0	−0.5	−8.0
240	15	−0.5	−7.50	−0.5	−7.5	1	15	−0.5	−7.5
270	9	0	0	−1	−9	0	0	1.0	9.0
300	5	0.5	2.5	−0.5	−2.5	−1	−5	−0.5	−2.5
330	5	0.866	4.33	0.5	2.5	0	0	−0.5	−2.5
$\Sigma y = 189$		$\Sigma y\cos x = -22.76$		$\Sigma y\cos 2x = -14.0$		$\Sigma y\cos 3x = 5.0$		$\Sigma y\cos 4x = -3.0$	
$a_0 = 31.5$		$a_1 = -3.79$		$a_2 = -2.33$		$a_3 = 0.833$		$a_4 = -0.5$	

The coefficients of the sine terms can be found from a similar table.

$x°$	y	$\sin x$	$y\sin x$	$\sin 2x$	$y\sin 2x$	$\sin 3x$	$y\sin 3x$	$\sin 4x$	$y\sin 4x$
0	10	0	0	0	0	0	0	0	0
30	18	0.5	9	0.866	15.59	1	18	0.866	15.59
60	24	0.866	20.78	0.866	20.78	0	0	−0.866	−20.78
90	26	1	26	0	0	−1	−26	0	0
120	25	0.866	21.65	−0.866	−21.65	0	0	0.866	21.65
150	20	0.5	10	−0.866	−17.32	1	20	−0.866	−17.32
180	16	0	0	0	0	0	0	0	0
210	16	−0.5	−8	0.866	13.86	−1	−16	0.866	13.86
240	15	−0.866	−12.99	0.866	12.99	0	0	−0.866	−12.99
270	9	−1	−9	0	0	1	9	0	0
300	5	−0.866	−4.33	−0.866	−4.33	0	0	0.866	4.33
330	5	−0.5	−2.5	−0.866	−4.33	−1	−5	−0.866	−4.33
		$\Sigma y\sin x = 50.61$		$\Sigma y\sin 2x = 15.59$		$\Sigma y\sin 3x = 0$		$\Sigma y\sin 4x = 0.01$	
		$b_1 = 8.44$		$b_2 = 2.60$		$b_3 = 0$		$b_4 = 0.002$	

$$\therefore\ f(x) = 15.8 - 3.79\cos x - 2.33\cos 2x + 0.833\cos 3x - 0.50\cos 4x + \ldots$$
$$+ 8.44\sin x + 2.60\sin 2x + \quad 0 \quad + 0.002\sin 4x + \ldots$$

Note that corresponding sine and cosine terms can be combined as compound sine terms as follows

(a) $8.44 \sin x - 3.79 \cos x = A_1 \sin(x - \alpha_1)$

$$A_1^2 = 8.44^2 + 3.79^2 = 71.23 + 14.36 = 85.59 \qquad \therefore A_1 = 9.25$$

$$\tan \alpha_1 = \frac{3.79}{8.44} = 0.4491 \qquad \therefore \alpha_1 = 24°11'$$

$$\therefore 8.44 \sin x - 3.79 \cos x = 9.25 \sin(x - 24°11')$$

(b) $2.60 \sin 2x - 2.33 \cos 2x = A_2 \sin(2x - \alpha_2)$

$$A_1^2 = 2.60^2 + 2.33^2 = 6.76 + 5.43 = 12.19 \qquad \therefore A_2 = 3.49$$

$$\tan \alpha_2 = \frac{2.33}{2.60} = 0.8962 \qquad \therefore \alpha_2 = 41°52'$$

$$\therefore 2.60 \sin 2x - 2.33 \cos 2x = 3.49 \sin(2x - 41°52')$$

(c) $0.833 \cos 3x = 0.833 \sin(3x + 90°)$

(d) $0.002 \sin 4x - 0.50 \cos 4x = A_4 \sin(4x - \alpha_4)$

$$A_4^2 = 0.002^2 + 0.50^2 = 0.0000 + 0.25 \qquad \therefore A_4 = 0.50$$

$$\tan \alpha_4 = \frac{0.05}{0.002} = 250 \qquad \therefore \alpha_4 = 89°46'$$

$$\therefore 0.002 \sin 4x - 0.50 \cos 4x = 0.50 \sin(4x - 89°46')$$

$$\therefore \underline{f(x) = 15.8 + 9.25 \sin(x - 24°11') + 3.49 \sin(2x - 41°52')}$$
$$\underline{+ 0.833 \sin(3x + 90°) + 0.50 \sin(4x - 89°46') + \ldots}$$

From this result, we can see that

(a) the amplitudes of succeeding harmonics decrease in value. This is not always so, but in practice more than five harmonics need seldom be calculated.

(b) the first term 15.8 (i.e. $\frac{1}{2}a_0$) is a constant term and, in graphical terms, raises the whole waveform 15.8 units on the y-scale. In electrical problems, this represents a d.c. component of the oscillatory current or voltage.

Therefore, for this example,

15.8	is the constant term
$9.25 \sin(x - 24°11')$	is the first harmonic or fundamental
$3.49 \sin(2x - 41°52')$	is the second harmonic
$0.833 \sin(3x + 90°)$	is the third harmonic
$0.50 \sin(4x - 89°46')$	is the fourth harmonic

Finally, we are required on occasions to express the amplitude of a particular harmonic as a percentage of the fundamental. For example, in this case, the

$$\text{percentage third harmonic} = \frac{0.833}{9.25} \times 100\% = \underline{9.01\%.}$$

Success of the series obtained If we calculate the values of $F(x)$ using the five terms of the series above, we can compare the results with the function values given originally in the example.

$x°$	0	30	60	90	120	150	180	210	240	270	300	330
$f(x)$	10	18	24	26	25	20	16	16	15	9	5	5
$F(x)$	10.01	18.08	24.04	26.07	25.00	20.13	15.93	16.20	14.88	9.19	4.93	5.13

Plotting the two graphs on the same axes will show how closely the first five terms of the calculated series represent the given function.

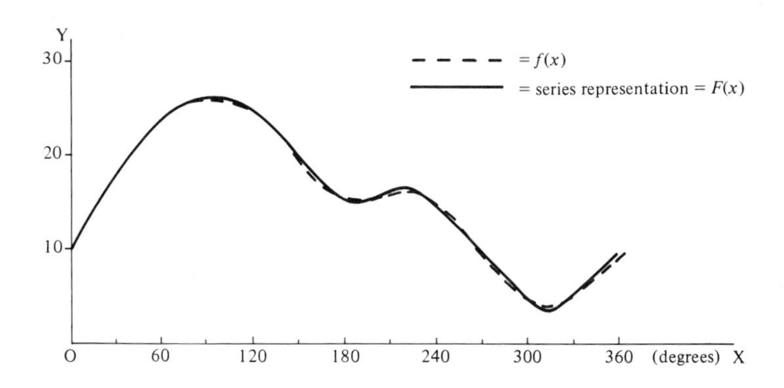

Example 2

One cycle of a periodic waveform $y = f(x)$ of period 2π is defined by the following set of function values:

$x°$	0	30	60	90	120	150	180	210	240	270	300	330
y	15	20	23	24	20	8	3	4	9	12	10	11

Determine the Fourier series up to and including the third harmonic.

x°	y	$\cos x$	$y \cos x$	$\cos 2x$	$y \cos 2x$	$\cos 3x$	$y \cos 3x$	$\sin x$	$y \sin x$	$\sin 2x$	$y \sin 2x$	$\sin 3x$	$y \sin 3x$
0	15	1	15	1	15	1	15	0	0	0	0	0	0
30	20	0.866	17.32	0.5	10	0	0	0.5	10	0.866	17.32	1	20
60	23	0.5	11.5	−0.5	−11.5	−1	−23	0.866	19.92	0.866	19.92	0	0
90	24	0	0	−1	−24	0	0	1	24	0	0	−1	−24
120	20	−0.5	−10	−0.5	−10	1	20	0.866	17.32	−0.866	−17.32	0	0
150	8	−0.866	−6.93	0.5	4	0	0	0.5	4	−0.866	−6.93	1	8
180	3	−1	−3	1	3	−1	−3	0	0	0	0	0	0
210	4	−0.866	−3.46	0.5	2	0	0	−0.5	−2	0.866	3.46	−1	−4
240	9	−0.5	−4.5	−0.5	−4.5	1	9	−0.866	−7.79	0.866	7.79	0	0
270	12	0	0	−1	−12	0	0	−1	−12	0	0	1	12
300	10	0.5	5	−0.5	−5	−1	−10	−0.866	−8.66	−0.866	−8.66	0	0
330	11	0.866	9.53	0.5	5.5	0	0	−0.5	−5.5	−0.866	−9.53	−1	−11
$\Sigma y = 159$		$\Sigma y \cos x = 30.46$		$\Sigma y \cos 2x = -27.5$		$\Sigma y \cos 3x = 8.0$		$\Sigma y \sin x = 39.29$		$\Sigma y \sin 2x = 6.05$		$\Sigma y \sin 3x = 1.00$	
$a_0 = 26.5$		$a_1 = 5.08$		$a_2 = -4.58$		$a_3 = 1.33$		$b_1 = 6.55$		$b_2 = 1.01$		$b_3 = 0.17$	

$$\therefore\ f(x) = \tfrac{1}{2}(26.5) + 5.08\cos x - 4.58\cos 2x + 1.35\cos 3x + \ldots$$
$$+ 6.55\sin x + 1.01\sin 2x + 0.17\sin 3x + \ldots$$

Expressing the results as compound sines, we have

(a) $6.55\sin x + 5.08\cos x = A_1 \sin(x + \alpha_1)$

$$A_1^2 = 6.55^2 + 5.08^2 = 42.90 + 25.81 = 68.71 \qquad \therefore\ A_1 = 8.29$$
$$\tan\alpha_1 = \frac{5.08}{6.55} = 0.7756 \qquad \therefore\ \alpha_1 = 37°48'$$
$$\therefore\ 6.55\sin x + 5.08\cos x = 8.29\sin(x + 37°48')$$

(b) $1.01\sin 2x - 4.58\cos 2x = A_2\sin(2x - \alpha_2)$

$$A_2^2 = 1.01^2 + 4.58^2 = 1.02 + 20.98 = 22.00 \qquad \therefore\ A_2 = 4.69$$
$$\tan\alpha_2 = \frac{4.58}{1.01} = 4.5347 \qquad \therefore\ \alpha_2 = 77°34'$$
$$\therefore\ 1.01\sin 2x - 4.58\cos 2x = 4.69\sin(2x - 77°34')$$

(c) $0.17\sin 3x + 1.33\cos 3x = A_3\sin(3x + \alpha_3)$

$$A_3^2 = 0.17^2 + 1.33^2 = 0.0289 + 1.769 = 1.7978 \qquad A_3 = 1.34$$
$$\tan\alpha_3 = \frac{1.33}{0.17} = 7.8235 \qquad \therefore\ \alpha_3 = 82°43'$$
$$\therefore\ 0.17\sin 3x + 1.33\cos 3x = 1.34\sin(3x + 82°43')$$

$$\therefore\ \underline{f(x) = 13.25 + 8.29\sin(x + 37°48') + 4.69\sin(2x - 77°34')}$$
$$\underline{+ 1.34\sin(3x + 82°43') + \ldots}$$

Function values $F(x)$ calculated from these four terms of the series can be compared with the corresponding function values $f(x)$ given in the original data.

$x°$	0	30	60	90	120	150	180	210	240	270	300	330
$f(x)$	15	20	23	24	20	8	3	4	9	12	10	11
$F(x)$	15.08	19.68	23.30	24.21	19.13	9.14	2.26	3.99	9.53	11.45	10.21	11.05

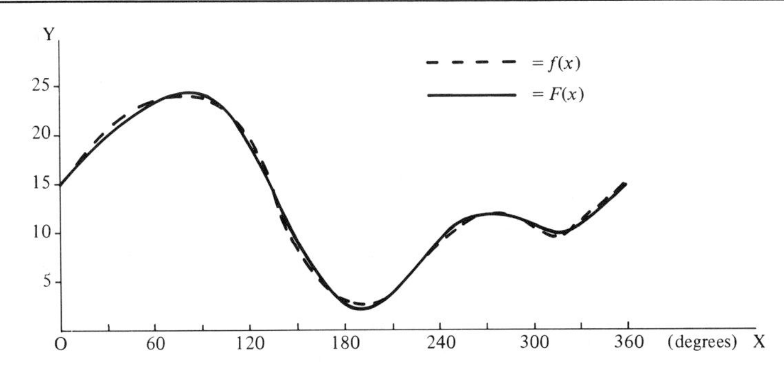

Example 3

A periodic function of period 2π is represented by the following table of function values extending over one cycle:

$x°$	0	30	60	90	120	150	180	210	240	270	300	330
$f(x)$	30	25	14	6	4	7	13	7	4	6	14	25

If we sketch the graph, we have

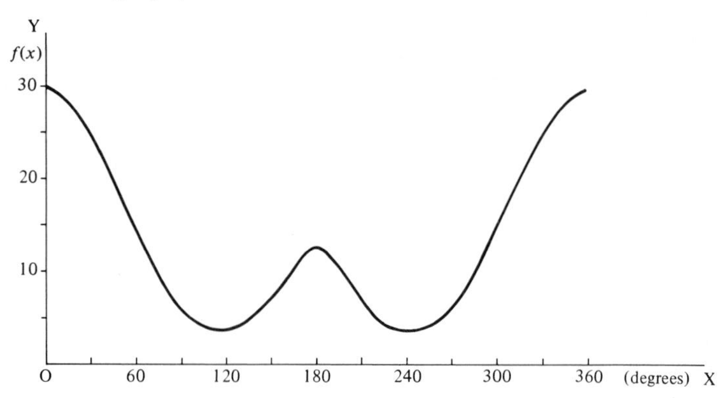

In this case, $f(x) = f(x+2\pi)$

Also $f(x) = f(-x)$

The function is therefore an *even* function and contains cosine terms only.

$$\therefore\ f(x) = \tfrac{1}{2}a_0 + \sum_{n=1}^{\infty} f(x)\cos nx$$

$x°$	y	$\cos x$	$y\cos x$	$\cos 2x$	$y\cos 2x$	$\cos 3x$	$y\cos 3x$
0	30	1	30	1	30	1	30
30	25	0.866	21.65	0.5	12.5	0	0
60	14	0.5	7	−0.5	−7	−1	−14
90	6	0	0	−1	−6	0	0
120	4	−0.5	−2	−0.5	−2	1	4
150	7	−0.866	−6.06	0.5	3.5	0	0
180	13	−1	−13	1	13	−1	−13
210	7	−0.866	−6.06	0.5	3.5	0	0
240	4	−0.5	−2	−0.5	−2	1	4
270	6	0	0	−1	−6	0	0
300	14	0.5	7	−0.5	−7	−1	−14
330	25	0.866	21.65	0.5	12.5	0	0
$\Sigma y = 155$		$\Sigma y\cos x = 58.18$		$\Sigma y\cos 2x = 45.0$		$\Sigma y\cos 3x = -3.0$	
$a_0 = 25.83$		$a_1 = 9.70$		$a_2 = 7.50$		$a_3 = -0.50$	

$$\therefore\ \underline{f(x) = 12.9 + 9.70\cos x + 7.50\cos 2x - 0.50\cos 3x + \ldots}$$

For interest, we can compare calculated function values $F(x)$ using these first four terms of the series, with the original given function values $f(x)$.

x°	0	30	60	90	120	150	180	210	240	270	300	330
$f(x)$	30	25	14	6	4	7	13	7	4	6	14	25
$F(x)$	29.6	25.1	14.5	5.4	3.8	8.3	11.2	8.3	3.8	5.4	14.5	25.1

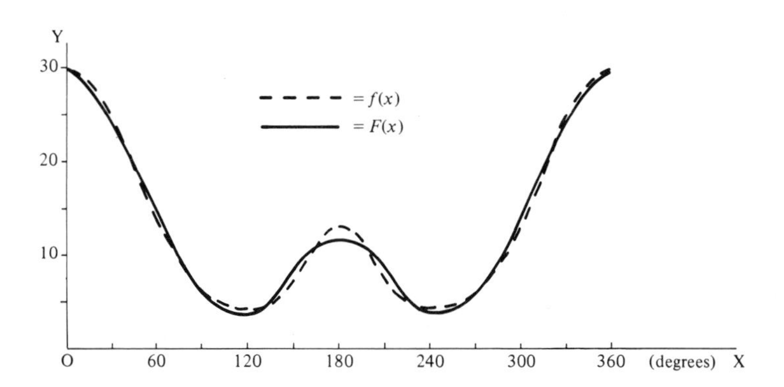

Example 4

Values of the ordinates for one cycle of a periodic waveform of period 2π are given in the table below:

x°	0	30	60	90	120	150	180	210	240	270	300	330
y	0	3	9	17	16	2	0	−3	−9	−17	−16	−2

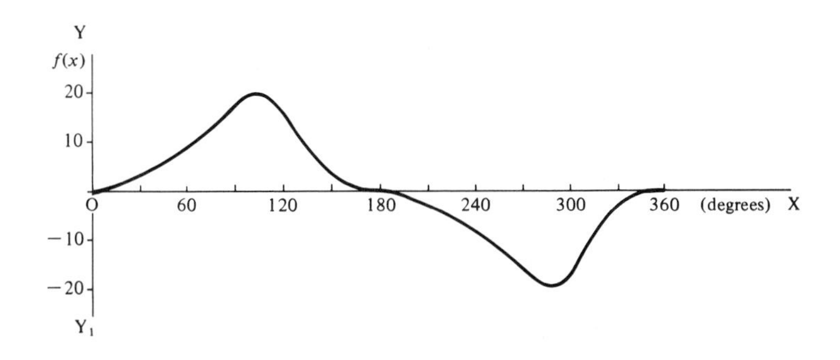

In this example, $f(x) = f(x + 2\pi)$ and also we see that $f(x) = -f(x + \pi)$, i.e. the function contains odd harmonics only.

x°	y	$\cos x$	$y \cos x$	$\cos 3x$	$y \cos 3x$	$\cos 5x$	$y \cos 5x$	$\sin x$	$y \sin x$	$\sin 3x$	$y \sin 3x$	$\sin 5x$	$y \sin 5x$
0	0	1	0	1	0	1	0	0	0	0	0	0	0
30	3	0.866	2.598	0	0	−0.866	−2.598	0.5	1.5	1	3	0.5	1.5
60	9	0.5	4.5	−1	−9	0.5	4.5	0.866	7.794	0	0	−0.866	−7.794
90	17	0	0	0	0	0	0	1	17	−1	−17	1	17
120	16	−0.5	−8.0	1	16	−0.5	−8	0.866	13.856	0	0	−0.866	−13.856
150	2	−0.866	−1.732	0	0	0.866	1.732	0.5	1	1	2	0.5	1
180	0	−1	0	−1	0	−1	0	0	0	0	0	0	0
210	−3	−0.866	2.598	0	0	0.866	−2.598	−0.5	1.5	−1	3	−0.5	1.5
240	−9	−0.5	4.5	1	−9	−0.5	4.5	−0.866	7.794	0	0	0.866	−7.794
270	−17	0	0	0	0	0	0	−1	17	1	−17	−1	17
300	−16	0.5	−8	−1	16	0.5	−8	−0.866	13.856	0	0	0.866	−13.856
330	−2	0.866	−1.732	0	0	−0.866	1.732	−0.5	1	−1	2	−0.5	1
$\Sigma y = 0$		$\Sigma y \cos x = -5.268$		$\Sigma y \cos 3x = 14$		$\Sigma y \cos 5x = -8.732$		$\Sigma y \sin x = 82.3$		$\Sigma y \sin 3x = -24$		$\Sigma y \sin 5x = -4.3$	
$a_0 = 0$		$a_1 = -0.878$		$a_3 = 2.33$		$a_5 = -1.46$		$b_1 = 13.72$		$b_3 = -4.0$		$b_5 = -0.717$	

$$\therefore\ f(x) = -0.878\cos x + 2.33\cos 3x - 1.46\cos 5x + \ldots$$
$$+ 13.7\sin x - 4.0\sin 3x - 0.717\sin 5x + \ldots$$

Expressing the corresponding sine and cosine terms as compound sine terms, as before, we get

$$13.7\sin x - 0.878\cos x = 13.7\sin(x - 3°40')$$
$$4.0\sin 3x - 2.33\cos 3x = 4.63\sin(3x - 30°13')$$
$$0.717\sin 5x + 1.46\cos 5x = 1.63\sin(5x + 63°51')$$
$$\therefore\ \underline{f(x) = 13.7\sin(x - 3°40') - 4.63\sin(3x - 30°13')}$$
$$\underline{-1.63\sin(5x + 63°51') + \ldots}$$

Example 5

A periodic voltage waveform $v = f(t)$ volts of period 6 ms is shown.

(a) Express the function as a Fourier series up to and including the third harmonic.

(b) Determine the percentage third harmonic.

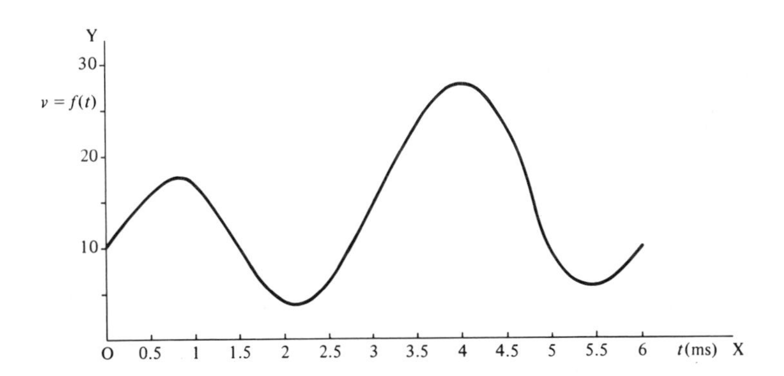

t (ms)	0	0.5	1.0	1.5	2.0	2.5	3.0	3.5	4.0	4.5	5.0	5.5	6.0
v (volts)	10	16	17	10	4	6	15	24	28	23	9	6	10

$$v = f(t) = f(t+6) \qquad \therefore\ \text{Period } T = 6 \text{ ms}$$

The first step is to convert the periodic interval into 2π by changing the units.

Put $x = \dfrac{2\pi t}{T} = \dfrac{2\pi t}{6}$, i.e. $x = \dfrac{\pi t}{3}$.

When $t = 0$, $x = 0$

$t = 6$, $x = 2\pi$

The table of values can now be re-written as follows

t	0	0.5	1.0	1.5	2.0	2.5	3.0	3.5	4.0	4.5	5.0	5.5
x	0	$\frac{\pi}{6}$	$\frac{\pi}{3}$	$\frac{\pi}{2}$	$\frac{2\pi}{3}$	$\frac{5\pi}{6}$	π	$\frac{7\pi}{6}$	$\frac{4\pi}{3}$	$\frac{3\pi}{2}$	$\frac{5\pi}{3}$	$\frac{11\pi}{6}$
	0	30	60	90	120	150	180	210	240	270	300	330
v	10	16	17	10	4	6	15	24	28	23	9	6

We then proceed as in previous examples, in this case using values of v as y, and obtain the series for $y = F(x)$. Substitution of the original units then gives the required result $v = f(t)$.

x°	y	$\cos x$	$y \cos x$	$\cos 2x$	$y \cos 2x$	$\cos 3x$	$y \cos 3x$	$\sin x$	$y \sin x$	$\sin 2x$	$y \sin 2x$	$\sin 3x$	$y \sin 3x$
0	10	1	10	1	10	1	10	0	0	0	0	0	0
30	16	0.866	13.86	0.5	8	0	0	0.5	8	0.866	13.86	1	16
60	17	0.5	8.5	−0.5	−8.5	−1	−17	0.866	14.72	0.866	14.72	0	0
90	10	0	0	−1	−10	0	0	1	10	0	0	−1	−10
120	4	−0.5	−2	−0.5	−2	1	4	0.866	3.464	−0.866	−3.464	0	0
150	6	−0.866	−5.196	0.5	3	0	0	0.5	3	−0.866	−5.196	1	6
180	15	−1	−15	1	15	−1	−15	0	0	0	0	0	0
210	24	−0.866	−20.78	0.5	12	0	0	−0.5	−12	0.866	20.78	−1	−24
240	28	−0.5	−14	−0.5	−14	1	28	−0.866	−24.25	0.866	24.25	0	0
270	23	0	0	−1	−23	0	0	−1	−23	0	0	1	23
300	9	0.5	4.5	−0.5	−4.5	−1	−9	−0.866	−7.794	−0.866	−7.794	0	0
330	6	0.866	5.196	0.5	3	0	0	−0.5	−3	−0.866	−5.196	−1	−6
$\Sigma y = 168$		$\Sigma y \cos x = -14.92$		$\Sigma y \cos 2x = -11.0$		$\Sigma y \cos 3x = 1.0$		$\Sigma y \sin x = -15.27$		$\Sigma y \sin 2x = 51.96$		$\Sigma y \sin 3x = 5.0$	
$a_0 = 28.0$		$a_1 = -2.49$		$a_2 = -1.83$		$a_3 = 0.167$		$b_1 = -2.55$		$b_2 = 8.66$		$b_3 = 0.833$	

$$\therefore\ F(x) = 14.0 - 2.49\cos x - 1.83\cos 2x + 0.167\cos 3x + \ldots$$
$$- 2.55\sin x + 8.66\sin 2x + 0.833\sin 3x + \ldots$$

We can now revert back to t as the variable, since we previously established that $x = \dfrac{\pi t}{3}$, i.e. $x = \omega t$ where $\omega = \dfrac{\pi}{3}$.

$$\therefore\ f(t) = 14.0 - 2.49\cos\omega t - 1.83\cos 2\omega t + 0.167\cos 3\omega t + \ldots$$
$$- 2.55\sin\omega t + 8.66\sin 2\omega t + 0.833\sin 3\omega t + \ldots$$

(a) $2.55\sin\omega t + 2.49\cos\omega t = A_1\sin(\omega t + \alpha_1)$

$$A_1^2 = 2.55^2 + 2.49^2 = 6.503 + 6.200 = 12.703 \qquad \therefore\ A_1 = 3.564$$

$$\tan\alpha_1 = \frac{2.49}{2.55} = 0.9765 \qquad \therefore\ \alpha_1 = 44^\circ 19' = 0.7735 \text{ radians}$$

$$\therefore\ 2.55\sin\omega t + 2.49\cos\omega t = 3.564\sin(\omega t + 0.774)$$

(b) $8.66\sin 2\omega t - 1.83\cos 2\omega t = A_2\sin(2\omega t - \alpha_2)$

$$A_2^2 = 8.66^2 + 1.83^2 = 74.996 + 3.349 = 78.345 \qquad \therefore\ A_2 = 8.851$$

$$\tan\alpha_2 = \frac{1.83}{8.66} = 0.2113 \qquad \therefore\ \alpha_2 = 11^\circ 56' = 0.2083 \text{ radians}$$

$$\therefore\ 8.66\sin 2\omega t - 1.83\cos 2\omega t = 8.851\sin(2\omega t - 0.208)$$

(c) $0.833\sin 3\omega t + 0.167\cos 3\omega t = A_3\sin(3\omega t + \alpha_3)$

$$A_3^2 = 0.833^2 + 0.167^2 = 0.6939 + 0.0279 = 0.7218 \quad \therefore\ A_3 = 0.8495$$

$$\tan\alpha_3 = \frac{0.167}{0.833} = 0.2005 \qquad \therefore\ \alpha_3 = 11^\circ 20' = 0.1979 \text{ radians}$$

$$\therefore\ 0.833\sin 3\omega t + 0.167\cos 3\omega t = 0.8495\sin(3\omega t + 0.198)$$

$$\therefore\ \underline{v = 14.0 - 3.56\sin(\omega t + 0.774) + 8.85\sin(2\omega t - 0.208)}$$
$$\underline{+ 0.850\sin(3\omega t + 0.198) + \ldots}$$

where $\omega = \dfrac{\pi}{3}$.

$$\text{Percentage third harmonic} = \frac{0.850}{3.56} \times 100\% = 23.876\%, \quad \text{i.e. } \underline{23.9\%}$$

Exercise 19

The periodic functions shown in Questions 1 to 7 are each of period 2π. In each case, express the function as a Fourier series up to and including the third harmonic.

1.

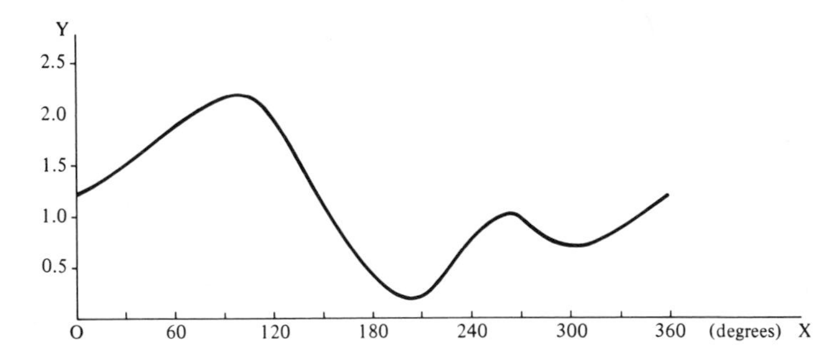

$x°$	0	30	60	90	120	150	180	210	240	270	300	330
y	1.2	1.5	1.9	2.2	2.0	1.1	0.4	0.2	0.8	1.0	0.7	0.9

2.

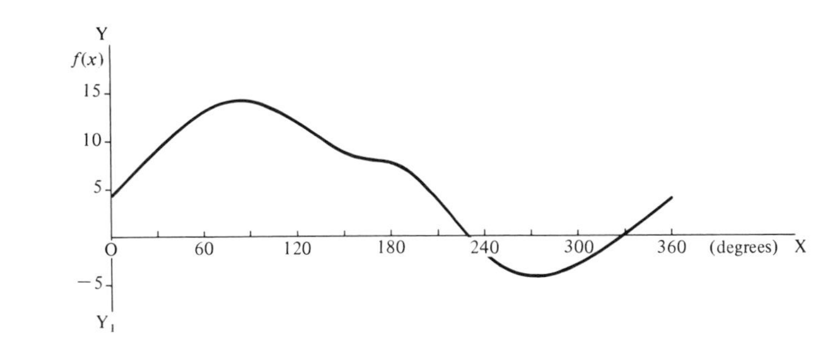

$x°$	0	30	60	90	120	150	180	210	240	270	300	330
y	4.0	8.7	13.0	14.0	12.0	9.0	8.0	4.0	−2.0	−4.0	−3.0	0

3.

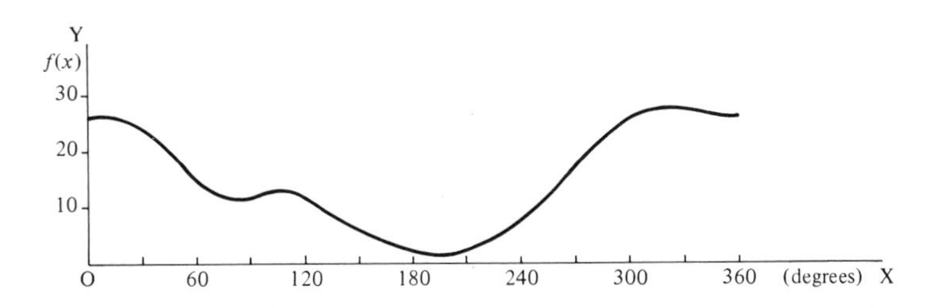

$x°$	0	30	60	90	120	150	180	210	240	270	300	330
y	26	24	14	12	12	6	2	2	8	18	26	28

4.

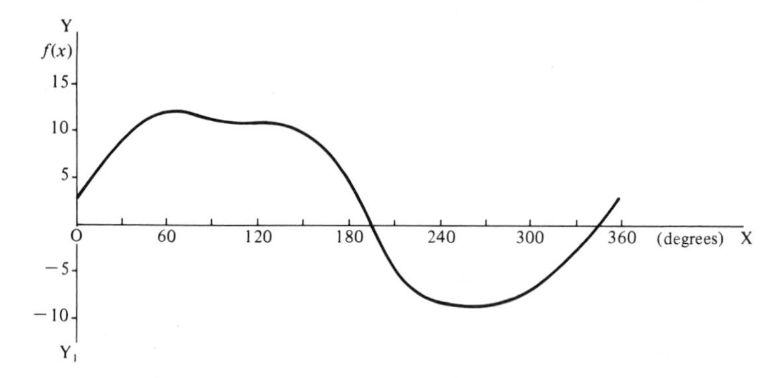

$x°$	0	30	60	90	120	150	180	210	240	270	300	330
y	2.8	9.0	12.0	11.5	10.8	10.0	5.0	−5.0	−8.0	−8.5	−7.5	−3.0

5.

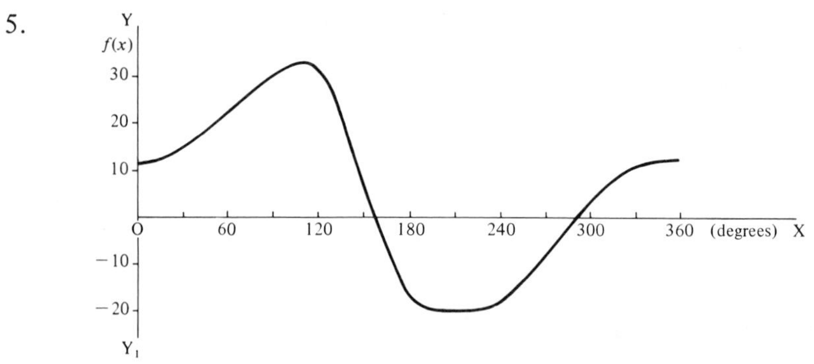

$x°$	0	30	60	90	120	150	180	210	240	270	300	330
y	12	15	22	30	32	8	−17	−20	−18	−8	4	11

6.

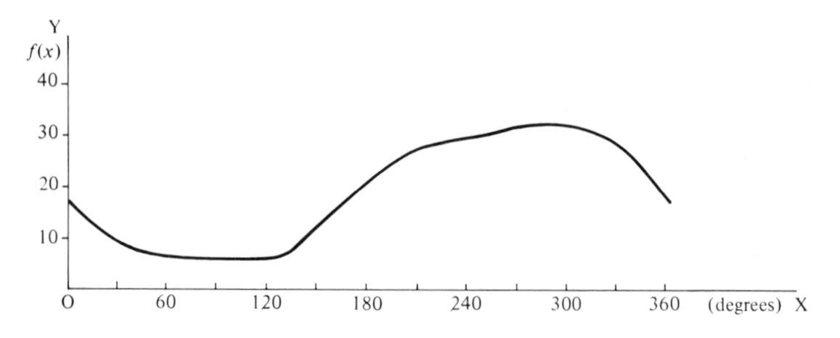

$x°$	0	30	60	90	120	150	180	210	240	270	300	330
y	17.5	9.4	7.0	6.0	5.5	12.0	21.5	27.8	30.0	32.0	32.5	28.5

7.

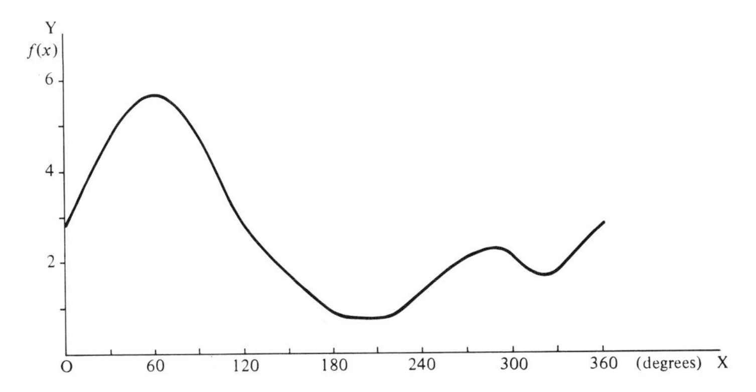

$x°$	0	30	60	90	120	150	180	210	240	270	300	330
y	2.75	4.76	5.77	4.75	2.93	1.64	0.95	0.78	1.27	2.15	2.13	1.82

8. A periodic voltage $v = f(x)$ is defined over one cycle by the waveform shown. Determine the first three terms of the Fourier series representing the function.

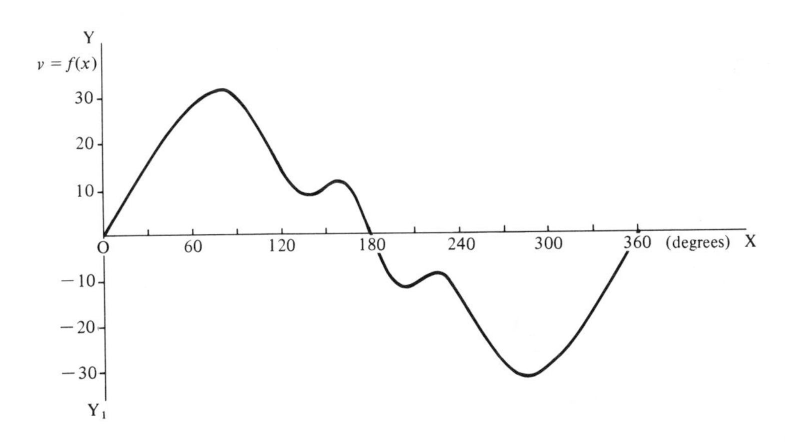

$x°$	0	30	60	90	120	150	180	210	240	270	300	330
v	0	17	29	30	14	11	0	−11	−14	−30	−29	−17

9. The function $y = f(u)$ is defined over the range $u = 0$ to $u = 60$ by the following table of function values. Also $f(u) = f(u + 60)$.

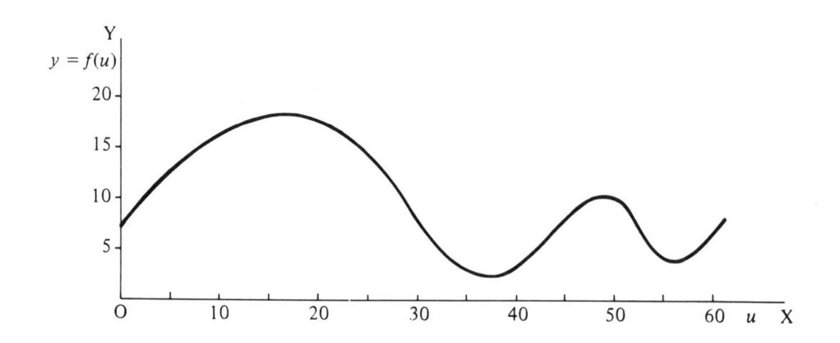

u	0	5	10	15	20	25	30	35	40	45	50	55	60
y	7.0	12.5	16.3	18.2	18.0	15.0	8.2	3.0	3.5	8.5	10.2	4.4	7.0

(a) Determine the Fourier series up to the third harmonic to represent the function.

(b) Calculate the percentage second harmonic.

10. A periodic current $i = f(t)$ amperes of period 18 ms has the waveform shown.

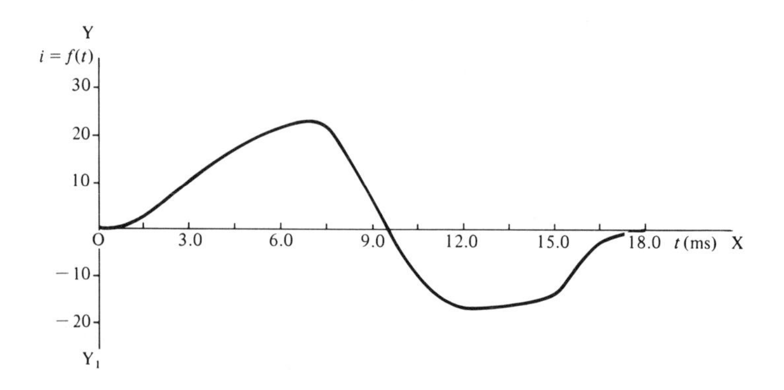

t (ms)	0	1.5	3.0	4.5	6.0	7.5	9.0	10.5	12.0	13.5	15.0	16.5	18.0
i (amperes)	0	3	10	17	22	22	7	−11	−16	−16	−14	−2	0

Determine the Fourier series up to and including the third harmonic.

6.3 REVISION SUMMARY

1. Fourier series coefficients

$$f(x) = \tfrac{1}{2}a_0 + \sum_{n=1}^{\infty} \{a_n \cos nx + b_n \sin nx\}$$

$$a_0 = \frac{1}{\pi}\int_0^{2\pi} f(x)\,\mathrm{d}x = 2 \times \text{mean value of } f(x) \text{ over a period}$$

$$a_n = \frac{1}{\pi}\int_0^{2\pi} f(x)\cos nx\,\mathrm{d}x = 2 \times \text{mean value of } f(x)\cos nx \text{ over a period}$$

$$b_n = \frac{1}{\pi}\int_0^{2\pi} f(x)\sin nx\,\mathrm{d}x = 2 \times \text{mean value of } f(x)\sin nx \text{ over a period}$$

2. **Trapezoidal rule** (for approximate integration)

$$I \approx s\{y_0 + y_1 + y_2 + \ldots + y_{k-1}\}$$

where s = width of each strip

k = number of strips

Chapter 7

APPLICATIONS OF FOURIER SERIES

7.1 HALF-WAVE RECTIFIER OUTPUT

7.1.1 Period 2π

If an alternating current $i = A\sin x$ is passed through a half-wave rectifier, the negative half of each cycle is suppressed.

Input *Output*

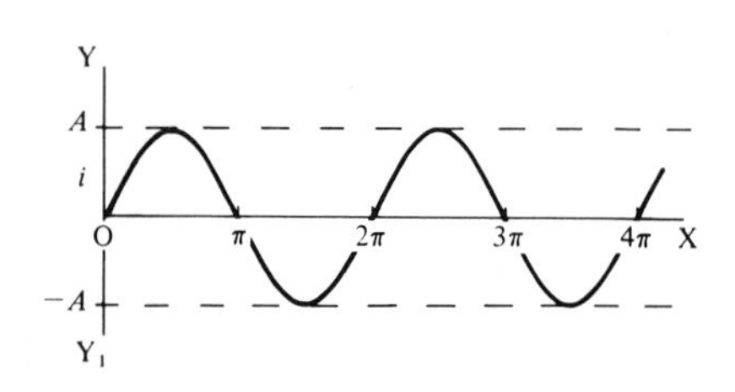

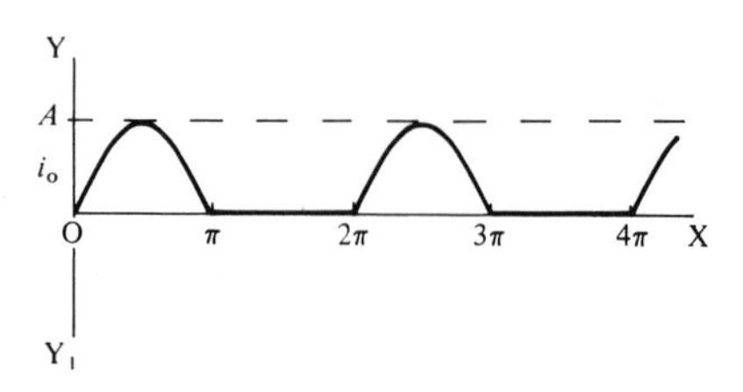

The output function can therefore be defined as

$$i = A\sin x \qquad 0 < x < \pi$$

$$i = 0 \qquad \pi < x < 2\pi$$

$$i = f(x) = f(x+2\pi)$$

Now $\quad i = \tfrac{1}{2}a_0 + \sum_{n=1}^{\infty} \{a_n \cos nx + b_n \sin nx\}$

where $\quad a_0 = \dfrac{1}{\pi}\displaystyle\int_0^{2\pi} f(x)\,dx = \dfrac{1}{\pi}\int_0^{\pi} A\sin x\,dx = \dfrac{A}{\pi}\Big[-\cos x\Big]_0^{\pi}$

$$= \frac{A}{\pi}\{-\cos\pi + 1\} = \frac{2A}{\pi} \qquad \therefore\ \underline{a_0 = \frac{2A}{\pi}}$$

$$a_n = \frac{1}{\pi}\int_0^{2\pi} f(x)\cos nx\,dx = \frac{1}{\pi}\int_0^{\pi} A\sin x\cos nx\,dx$$

$$= \frac{A}{2\pi}\int_0^{\pi}\{\sin(n+1)x - \sin(n-1)x\}\,dx$$

$$= \frac{A}{2\pi}\left[\frac{-\cos(n+1)x}{n+1} + \frac{\cos(n-1)x}{n-1}\right]_0^\pi \qquad n \neq 1$$

$$= \frac{A}{2\pi}\left\{\frac{1}{n+1}(1-\cos[n+1]\pi) - \frac{1}{n-1}(1-\cos[n-1]\pi)\right\}$$

For n even, $(n+1)$ and $(n-1)$ are odd $\therefore \cos[n+1]\pi = \cos[n-1]\pi = -1$.

For n odd, $(n+1)$ and $(n-1)$ are even $\therefore \cos[n+1]\pi = \cos[n-1]\pi = 1$.

For n even $\quad a_n = \frac{A}{2\pi}\left\{\frac{2}{n+1} - \frac{2}{n-1}\right\} \qquad \therefore \underline{a_n = \frac{-2A}{\pi(n-1)(n+1)}}$

For n odd $\quad a_n = \frac{A}{2\pi}\{0-0\} \qquad \therefore \underline{a_n = 0} \quad n \neq 1$

When $n = 1$: $\quad a_1 = \frac{A}{2\pi}\int_0^\pi \sin(n+1)x\,dx = \frac{A}{2\pi}\int_0^\pi \sin 2x\,dx$

$$= \frac{A}{2\pi}\left[\frac{-\cos 2x}{2}\right]_0^\pi = 0 \qquad \therefore \underline{a_1 = 0}$$

$$b_n = \frac{1}{\pi}\int_0^{2\pi} f(x)\sin nx\,dx = \frac{1}{\pi}\int_0^\pi A\sin x\sin nx\,dx$$

$$= \frac{A}{2\pi}\int_0^\pi \{\cos(n-1)x - \cos(n+1)x\}dx$$

$$= \frac{A}{2\pi}\left[\frac{\sin(n-1)x}{n-1} - \frac{\sin(n+1)x}{n+1}\right]_0^\pi \qquad n \neq 1$$

$$= 0 \qquad \therefore \underline{b_n = 0} \qquad n \neq 1$$

For $n = 1$ $\quad b_1 = \frac{1}{\pi}\int_0^\pi A\sin x\sin x\,dx = \frac{A}{\pi}\int_0^\pi \sin^2 x\,dx$

$$= \frac{A}{\pi}\int_0^\pi \tfrac{1}{2}(1-\cos 2x)\,dx = \frac{A}{2\pi}\left[x - \frac{\sin 2x}{2}\right]_0^\pi$$

$$= \frac{A}{2\pi}\{\pi\} = \frac{A}{2} \qquad \therefore \underline{b_1 = \frac{A}{2}}$$

Therefore, we have: $\quad a_0 = \frac{2A}{\pi}$

$$a_n = \frac{-2A}{\pi(n-1)(n+1)} \quad n \text{ even}; \qquad a_n = 0 \quad n \text{ odd};$$

$$b_n = 0 \quad n \neq 1; \qquad b_1 = \frac{A}{2}$$

$$\therefore\ i = \frac{A}{\pi} + \frac{A}{2}\sin x - \frac{2A}{\pi}\left\{\frac{1}{1.3}\cos 2x + \frac{1}{3.5}\cos 4x + \frac{1}{5.7}\cos 6x + \ldots\right\}$$

$$i = \frac{A}{\pi}\left\{1 + \frac{\pi}{2}\sin x - 2\left(\frac{1}{1.3}\cos 2x + \frac{1}{3.5}\cos 4x + \frac{1}{5.7}\cos 6x + \ldots\right)\right\}$$

7.1.2 Period T

For a periodic function $i = f(t)$ of period T, this becomes

Input *Output*

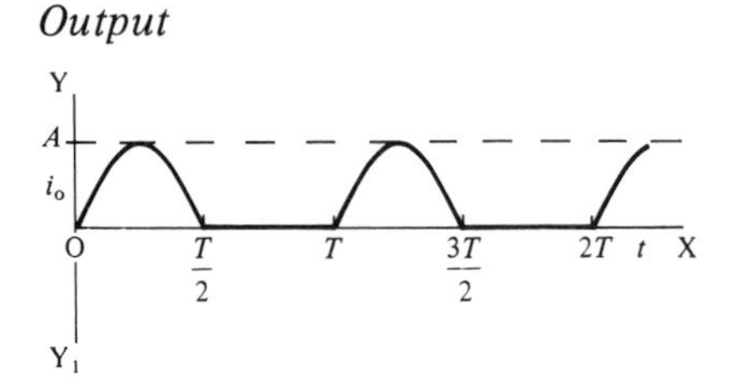

$$\omega = 2\pi f = \frac{2\pi}{T} \qquad \therefore\ T = \frac{2\pi}{\omega}$$

Put $\quad x = \dfrac{2\pi t}{T}$; $\quad$ when $t = 0,\ x = 0$

when $t = T,\ x = 2\pi$

The previous result then becomes

$$i = f(t) = \frac{A}{\pi} + \frac{A}{2}\sin\omega t - \frac{2A}{\pi}\left\{\frac{1}{1.3}\cos 2\omega t + \frac{1}{3.5}\cos 4\omega t + \ldots\right\}$$

$$i = \frac{A}{\pi}\left\{1 + \frac{\pi}{2}\sin\omega t - 2\left(\frac{1}{1.3}\cos 2\omega t + \frac{1}{3.5}\cos 4\omega t + \frac{1}{5.7}\cos 6\omega t + \ldots\right)\right\}$$

where $\omega = \dfrac{2\pi}{T}$.

7.2 FULL-WAVE RECTIFIER OUTPUT

7.2.1 Period 2π

In this case, the output from an input current $i = A\sin x$ is as shown.

Input *Output*

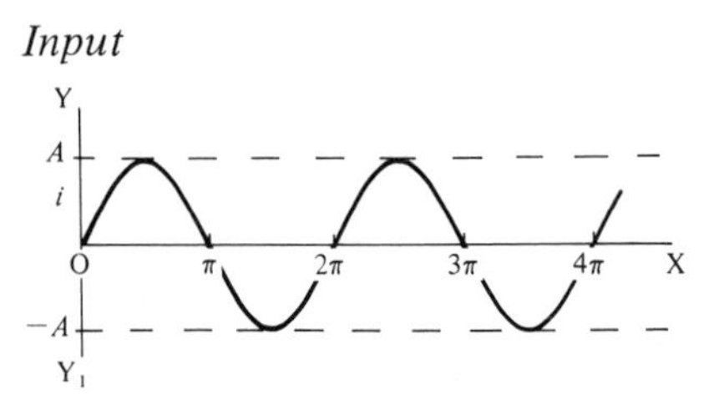

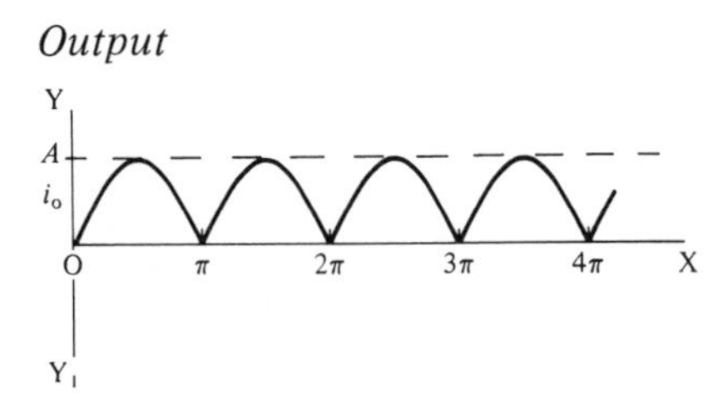

The output function may therefore be regarded as

$$f(x) = A\sin x \qquad 0 < x < \pi$$
$$f(x) = -A\sin x \qquad \pi < x < 2\pi$$
$$f(x) = f(x+2\pi)$$

Also, we note that $f(x) = f(x+\pi)$, $\therefore$ series contains *even* harmonics only.

$$a_0 = \frac{1}{\pi}\int_{-\pi}^{\pi} f(x)\,dx = \frac{2}{\pi}\int_0^{\pi} A\sin x\,dx = \frac{2A}{\pi}\Big[-\cos x\Big]_0^{\pi}$$

$$= \frac{2A}{\pi}\{1+1\} = \frac{4A}{\pi} \qquad \therefore\ a_0 = \frac{4A}{\pi}$$

$$a_n = \frac{2}{\pi}\int_0^{\pi} A\sin x\cos nx\,dx = \left(\frac{2A}{\pi}\right)\frac{1}{2}\int_0^{\pi}\{\sin(1+n)x + \sin(1-n)x\}dx$$

$$= \left(\frac{2A}{\pi}\right)\frac{1}{2}\int_0^{\pi}\{\sin(n+1)x - \sin(n-1)x\}dx$$

$$= \frac{A}{\pi}\left[\frac{-\cos(n+1)x}{n+1} + \frac{\cos(n-1)x}{n-1}\right]_0^{\pi} \qquad n \neq 1$$

$$= -\frac{A}{\pi}\left\{\frac{\cos(n+1)\pi}{n+1} - \frac{1}{n+1} - \frac{\cos(n-1)\pi}{n-1} + \frac{1}{n-1}\right\}$$

For n *even*, $(n+1)$ and $(n-1)$ are odd $\therefore \cos(n+1)\pi = \cos(n-1)\pi = -1$.

Then
$$a_n = -\frac{A}{\pi}\left\{-\frac{1}{n+1} - \frac{1}{n+1} + \frac{1}{n-1} + \frac{1}{n-1}\right\}$$

$$= \frac{2A}{\pi}\left\{\frac{1}{n+1} - \frac{1}{n-1}\right\} = \frac{-4A}{\pi(n-1)(n+1)}$$

$\therefore$ We have
$$a_0 = \frac{4A}{\pi}$$
$$a_n = \frac{-4A}{\pi(n-1)(n+1)} \qquad n \text{ even}$$
$$b_n = 0$$

The output is an even function.

$$\therefore\ i = \frac{2A}{\pi} - \frac{4A}{\pi}\left\{\frac{1}{1.3}\cos 2x + \frac{1}{3.5}\cos 4x + \frac{1}{5.7}\cos 6x + \ldots\right\}$$

i.e.
$$i = \frac{2A}{\pi}\left\{1 - 2\left(\frac{1}{1.3}\cos 2x + \frac{1}{3.5}\cos 4x + \frac{1}{5.7}\cos 6x + \ldots\right)\right\}$$

7.2.2 Period T

For an input of $i = A \sin \omega t$ of period T, $\omega = \dfrac{2\pi}{T}$.

Input *Output*

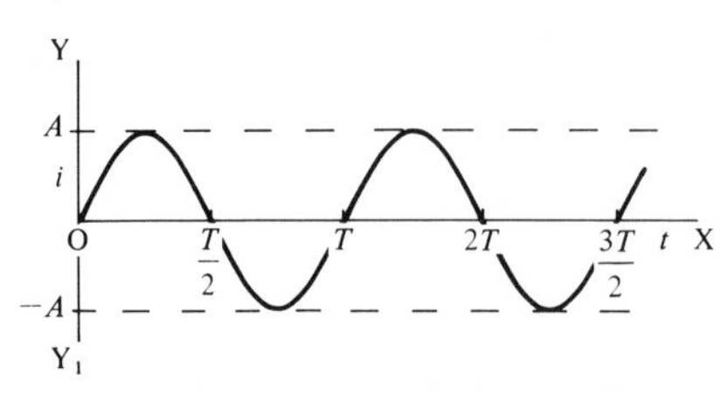

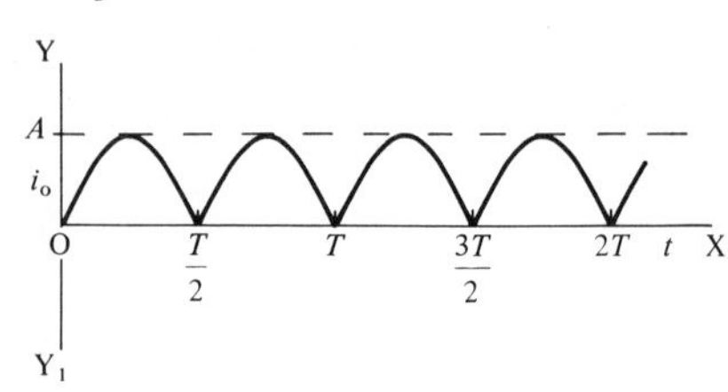

As in the previous example with period T, put $x = \dfrac{2\pi t}{T}$, so that

$$\text{when} \quad t = 0, \qquad x = 0$$
$$t = T, \qquad x = 2\pi$$

The result then becomes

$$i = \frac{2A}{\pi}\left\{1 - 2\left(\frac{1}{1.3}\cos 2\omega t + \frac{1}{3.5}\cos 4\omega t + \frac{1}{5.7}\cos 6\omega t + \ldots\right)\right\}$$

where $\omega = \dfrac{2\pi}{T}$.

7.3 PARSEVAL'S THEOREM

If a periodic function $y = f(x)$ of period $2L$ is defined over the interval $x = -L$ to $x = L$, then

$$f(x) = \tfrac{1}{2}a_0 + \sum_{n=1}^{\infty}\left\{a_n \cos\frac{n\pi x}{L} + b_n \sin\frac{n\pi x}{L}\right\} \tag{7.1}$$

If we multiply both sides by $f(x)$, we have

$$\{f(x)\}^2 = \tfrac{1}{2}a_0 f(x) + \sum_{n=1}^{\infty}\left\{a_n f(x) \cos\frac{n\pi x}{L} + b_n f(x) \sin\frac{n\pi x}{L}\right\}$$

Integrating both sides with respect to x from $-L$ to L gives

$$\int_{-L}^{L}\{f(x)\}^2\,dx = \tfrac{1}{2}a_0\int_{-L}^{L} f(x)\,dx + \sum_{n=1}^{\infty}\left\{a_n\int_{-L}^{L} f(x)\cos\frac{n\pi x}{L}\,dx + b_n\int_{-L}^{L} f(x)\sin\frac{n\pi x}{L}\,dx\right\} \tag{7.2}$$

From the definitions of the coefficients of the original series,

$$a_0 = \frac{1}{L}\int_{-L}^{L} f(x)\,dx \qquad \therefore\ La_0 = \int_{-L}^{L} f(x)\,dx$$

$\therefore$ In (7.2)

$$\tfrac{1}{2}a_0\int_{-L}^{L} f(x)\,dx = \tfrac{1}{2}a_0 \times La_0 = \frac{La_0^2}{2} \qquad (7.3)$$

Also $\quad a_n = \frac{1}{L}\int_{-L}^{L} f(x)\cos\frac{n\pi x}{L}\,dx \qquad \therefore\ La_n = \int_{-L}^{L} f(x)\cos\frac{n\pi x}{L}\,dx$

$\therefore$ In (7.2)

$$a_n\int_{-L}^{L} f(x)\cos\frac{n\pi x}{L}\,dx = a_n La_n = La_n^2 \qquad (7.4)$$

Similarly,

$$b_n = \frac{1}{L}\int_{-L}^{L} f(x)\sin\frac{n\pi x}{L}\,dx \qquad \therefore\ Lb_n = \int_{-L}^{L} f(x)\sin\frac{n\pi x}{L}\,dx$$

$\therefore$ In (7.2)

$$b_n\int_{-L}^{L} f(x)\sin\frac{n\pi x}{L}\,dx = b_n Lb_n = Lb_n^2 \qquad (7.5)$$

Substituting the results (7.3), (7.4) and (7.5) in (7.2), we obtain

$$\int_{-L}^{L}\{f(x)\}^2\,dx = \tfrac{1}{2}La_0^2 + \sum_{n=1}^{\infty}\{La_n^2 + Lb_n^2\}$$

$$\therefore\ \frac{1}{L}\int_{-L}^{L}\{f(x)\}^2\,dx = \tfrac{1}{2}a_0^2 + \sum_{n=1}^{\infty}\{a_n^2 + b_n^2\}$$

This is *Parseval's theorem*. Note that the right hand expression denotes $2 \times$ the mean value of $\{f(x)\}^2$ over a cycle.

For a periodic function $y = f(t)$ of period T defined in the interval $t = 0$ to $t = T$, the result becomes

$$\frac{2}{T}\int_{0}^{T}\{f(t)\}^2\,dt = \tfrac{1}{2}a_0^2 + \sum_{n=1}^{\infty}\{a_n^2 + b_n^2\}$$

7.4 RMS VALUE OF A PERIODIC CURRENT OR VOLTAGE EXPRESSED AS A FOURIER SERIES

The *root mean square* value of an alternating current $i = f(t)$ of period T is given by

$$(\text{RMS})^2 = \frac{1}{T}\int_{0}^{T} i^2\,dt$$

If i is expressed in the form of a Fourier series

$$i = f(t) = \tfrac{1}{2}a_0 + \sum_{n=1}^{\infty}\{a_n \cos n\omega t + b_n \sin n\omega t\} \qquad \text{where} \quad \omega = \frac{2\pi}{T}$$

Then, by Parseval's theorem,

$$\frac{2}{T}\int_0^T \{f(t)\}^2\,\mathrm{d}t = \tfrac{1}{2}a_0^2 + \sum_{n=1}^{\infty}\{a_n^2 + b_n^2\}$$

The left hand side is twice the square of the RMS value of $f(t)$.

$$\therefore\ (\text{RMS})^2 = \tfrac{1}{4}a_0^2 + \frac{1}{2}\sum_{n=1}^{\infty}\{a_n^2 + b_n^2\}$$

Finally, the RMS value is found by taking the square root of each side.

For a function having no d.c. component, $a_0 = 0$. Then

$$i = f(t) = \sum_{n=1}^{\infty}\{a_n \cos n\omega t + b_n \sin n\omega t\} \tag{7.6}$$

This can be written $\quad i = \sum_{n=1}^{\infty} I_n \sin(n\omega t + \phi_n)$

where $\quad I_n = \sqrt{a_n^2 + b_n^2}, \quad$ and $\quad \tan\phi_n = \dfrac{a_n}{b_n}.$

$$\therefore\ i = \sum_{n=1}^{\infty} I_n\{\sin n\omega t \cos\phi_n + \cos n\omega t \sin\phi_n\}$$

$$= \sum_{n=1}^{\infty}\{I_n \cos\phi_n \sin n\omega t + I_n \sin\phi_n \cos n\omega t\}$$

If we compare this result with (7.3)

$$a_n = I_n \sin\phi_n \qquad \text{and} \qquad b_n = I_n \cos\phi_n$$

$$\therefore\ (\text{RMS})^2 = \frac{1}{2}\sum_{n=1}^{\infty}\{a_n^2 + b_n^2\}$$

$$= \frac{1}{2}\sum_{n=1}^{\infty}\{I_n^2 \sin^2\phi_n + I_n^2 \cos^2\phi_n\} = \frac{1}{2}\sum_{n=1}^{\infty} I_n^2$$

$$\therefore\ \text{RMS} = \sqrt{\frac{1}{2}\{I_1^2 + I_2^2 + I_3^2 + \ldots + I_n^2 + \ldots\}}$$

7.5 MULTIPLICATION THEOREM

If $f(x)$ is a periodic function of period $2L$, defined over the interval $x = -L$ to $x = L$, then

$$f(x) = \tfrac{1}{2}a_0 + \sum_{n=1}^{\infty}\left\{a_n \cos\frac{n\pi x}{L} + b_n \sin\frac{n\pi x}{L}\right\} \tag{7.7}$$

Similarly, if $F(x)$ is a second periodic function of period $2L$ defined over the same interval,

$$F(x) = \tfrac{1}{2}A_0 + \sum_{n=1}^{\infty}\left\{A_n \cos\frac{n\pi x}{L} + B_n \sin\frac{n\pi x}{L}\right\} \tag{7.8}$$

Then (7.8) + (7.7) gives

$$F(x) + f(x) = \tfrac{1}{2}(A_0 + a_0) + \sum_{n=1}^{\infty}\left\{(A_n + a_n)\cos\frac{n\pi x}{L} + (B_n + b_n)\sin\frac{n\pi x}{L}\right\} \tag{7.9}$$

and (7.8) − (7.7) gives

$$F(x) - f(x) = \tfrac{1}{2}(A_0 - a_0) + \sum_{n=1}^{\infty}\left\{(A_n - a_n)\cos\frac{n\pi x}{L} + (B_n - b_n)\sin\frac{n\pi x}{L}\right\} \tag{7.10}$$

Applying Parseval's theorem to (7.9) and (7.10)

$$\frac{1}{L}\int_{-L}^{L}\{F(x) + f(x)\}^2\,dx = \tfrac{1}{2}(A_0 + a_0)^2 + \sum_{n=1}^{\infty}\{(A_n + a_n)^2 + (B_n + b_n)^2\} \tag{7.11}$$

$$\frac{1}{L}\int_{-L}^{L}\{F(x) - f(x)\}^2\,dx = \tfrac{1}{2}(A_n - a_0)^2 + \sum_{n=1}^{\infty}\{(A_n - a_n)^2 + (A_n - a_n)^2\} \tag{7.12}$$

Subtracting (7.12) from (7.11)

$$\frac{1}{L}\int_{-L}^{L} 4F(x)\ f(x)\,dx = \tfrac{1}{2}(4A_0 a_0) + \sum_{n=1}^{\infty}\{4A_n a_n + 4B_n b_n\}$$

$$\therefore\quad \frac{1}{L}\int_{-L}^{L} F(x)\ f(x)\,dx = \tfrac{1}{2}A_0 a_0 + \sum_{n=1}^{\infty}\{A_n a_n + B_n b_n\} \tag{7.13}$$

Note:

(a) Dividing both sides of result (7.13) by 2 gives

$$\frac{1}{2L}\int_{-L}^{L}\{F(x)\ f(x)\,dx = \tfrac{1}{4}A_0 a_0 + \frac{1}{2}\sum_{n=1}^{\infty}\{A_n a_n + B_n b_n\}$$

That is, the right hand side gives the average value of the product of the two functions over a complete period, i.e. $2L$

(b) When $F(x) = f(x)$, result (7.13) becomes

$$\frac{1}{L}\int_{-L}^{L} \{f(x)\}^2 \, dx = \tfrac{1}{2}a_0^2 + \sum_{n=1}^{\infty} \{a_n^2 + b_n^2\}$$

which we established earlier as Parseval's theorem.

7.6 POWER IN A CIRCUIT WHEN A VOLTAGE AND/OR CURRENT IS EXPRESSED AS A FOURIER SERIES

(a) If an alternating voltage $v = f(t)$ volts applied to a circuit results in a corresponding instantaneous current $i = F(t)$ amperes, where each function is of period T, then the average value of the power, vi watts, over one cycle can be obtained by use of the multiplication theorem.

If
$$v = f(t) = \tfrac{1}{2}a_0 + \sum_{n=1}^{\infty} \{a_n \cos n\omega t + b_n \sin n\omega t\}$$

and
$$i = F(t) = \tfrac{1}{2}A_0 + \sum_{n=1}^{\infty} \{A_n \cos n\omega t + B_n \sin n\omega t\}$$

then the average value of the power, p watts, over one cycle is given by

$$\frac{1}{T}\int_0^T f(t)\ F(t)\, dt = \frac{1}{4}A_0 a_0 + \frac{1}{2}\sum_{n=1}^{\infty} \{A_n a_n + B_n b_n\}$$

$$\therefore \text{ Average value of } vi = \frac{1}{4}A_0 a_0 + \frac{1}{2}\sum_{n=1}^{\infty} \{A_n a_n + B_n b_n\}$$

Example 1

If
$$v = 12.0 + 5.2 \cos \omega t + 2.4 \cos 2\omega t + 0.9 \cos 3\omega t + \ldots$$
$$+ 2.7 \sin \omega t + 1.8 \sin 2\omega t + 0.2 \sin 3\omega t + \ldots$$

and
$$i = 8.5 + 4.1 \cos \omega t + 2.0 \cos 2\omega t + 0.6 \cos 3\omega t + \ldots$$
$$+ 3.6 \sin \omega t + 1.2 \sin 2\omega t + 0.3 \sin 3\omega t + \ldots$$

So

$v = f(t)$	a_0	a_1	a_2	a_3	b_1	b_2	b_3
	12.0	5.2	2.4	0.9	2.7	1.8	0.2

$i = F(t)$	A_0	A_1	A_2	A_3	B_1	B_2	B_3
	8.5	4.1	2.0	0.6	3.6	1.2	0.3

$$\therefore \text{ Average } vi = \frac{1}{4}A_0 a_0 + \frac{1}{2}\sum_{n=1}^{\infty}\{A_n a_n + B_n b_n\}$$

$$= \frac{1}{4}(102) + \frac{1}{2}\{21.32 + 4.80 + 0.54 + 9.72 + 2.16 + 0.06\}$$

$$= 25.5 + \frac{1}{2}\{38.6\} = 25.5 + 19.3 = \underline{44.8\text{ W}}$$

(b) If a periodic voltage $v = f(t)$ is applied to a resistor of resistance R ohms, the instantaneous value of the power, p watts, dissipated is given by $p = \dfrac{v^2}{R}$.

Therefore, the average value of the power over one cycle (period T) becomes

$$\text{Average } p = \left(\frac{1}{R}\right)\frac{1}{T}\int_0^T \{f(t)\}^2\,dt$$

Example 2

If $R = 5.0\,\Omega$ and

$$v = f(t) = 8.6 + 5.4\cos\omega t + 1.8\cos 2\omega t - 0.3\cos 3\omega t + \ldots$$
$$+ 2.3\sin\omega t - 0.7\sin 2\omega t + 0.2\sin 3\omega t + \ldots$$

a_0	a_1	a_2	a_3	b_1	b_2	b_3
8.6	5.4	1.8	−0.3	2.3	−0.7	0.2

$$\therefore \text{ Average } p = \frac{1}{5}\left\{\frac{1}{4}(73.96) + \frac{1}{2}(29.16 + 3.24 + 0.09 + 5.29 + 0.49 + 0.04)\right\}$$

$$= \frac{1}{5}\left\{18.49 + \frac{1}{2}(38.31)\right\} = \frac{1}{5}\{18.49 + 19.16\} = \underline{7.53\text{ W}}$$

The expressions for the power stated earlier refer to a non-inductive circuit. Where a circuit also contains inductance, the results above would have to be multiplied by a power factor, a constant less than unity and denoted by $\cos\phi$ where $\cos\phi = \dfrac{R}{Z}$. R is the resistance and Z the impedance of the circuit in ohms.

7.7 SPECTRUM OF A WAVEFORM

For a periodic waveform $y = f(t)$ of period T

$$f(t) = \tfrac{1}{2}a_0 + \sum_{n=1}^{\infty}\{a_n\cos n\omega t + b_n\sin n\omega t\}$$

where $\omega = 2\pi f$

$n = 1, 2, 3, \ldots$

This can be written in the form

$$f(t) = \tfrac{1}{2}a_0 + \sum_{n=1}^{\infty}\{A_n \sin(n\omega t + \phi_n)\}$$

where $A_n = \sqrt{(a_n)^2 + (b_n)^2}$

$$\phi_n = \arctan \frac{a_n}{b_n}$$

If the amplitude A_n of each harmonic is plotted against the frequency, the result is a number of discrete ordinates which constitute the *spectrum* of the waveform.

Example

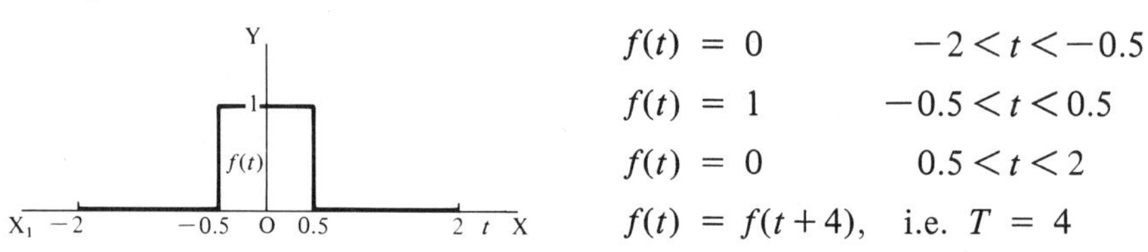

$f(t) = 0 \qquad -2 < t < -0.5$

$f(t) = 1 \qquad -0.5 < t < 0.5$

$f(t) = 0 \qquad 0.5 < t < 2$

$f(t) = f(t+4)$, i.e. $T = 4$

The function defined above is an even function and the Fourier series therefore contains no sine terms.

Determination of the Fourier series for the function in the usual way gives

$$f(t) = \frac{1}{4} + \frac{2}{\pi}\left\{\sum_{n=1}^{\infty} \frac{1}{n} \sin \frac{n\pi}{4} \cos n\omega t\right\}$$

where $\omega = 2\pi f = \dfrac{2\pi}{T} = \dfrac{\pi}{2}$.

The calculated amplitude A_n of each harmonic from $n = 1$ to $n = 32$ is then as follows

n	A_n	n	A_n	n	A_n	n	A_n
1	0.450	9	0.050	17	0.027	25	0.018
2	0.318	10	0.064	18	0.035	26	0.025
3	0.150	11	0.041	19	0.024	27	0.017
4	0	12	0	20	0	28	0
5	−0.090	13	−0.035	21	−0.022	29	−0.016
6	−0.106	14	−0.046	22	−0.029	30	−0.021
7	−0.064	15	−0.030	23	−0.020	31	−0.015
8	0	16	0	24	0	32	0

If the amplitude of each harmonic is plotted against a measure of the frequency (n) the spectrum obtained is as follows

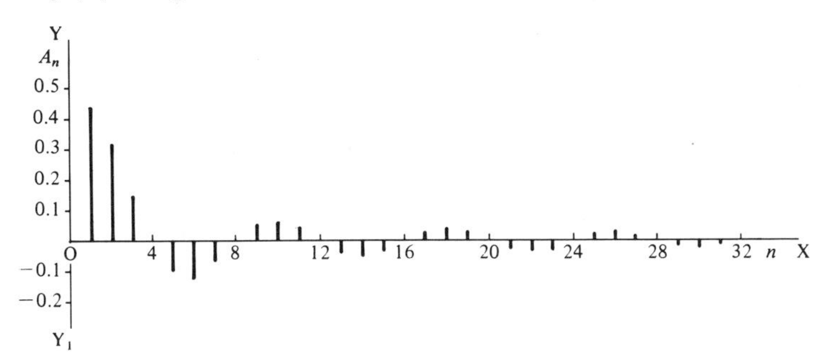

Note that the spectral lines occur at regular intervals of frequency $\dfrac{\omega}{2\pi}$.

The amplitudes of the higher harmonics are much reduced and the greater part of the information regarding the function is contained in the first few harmonics.

In some special cases where the period is made extremely large, the spectral lines occur much closer together and, as $T \to \infty$, the spectrum becomes a continuous curve.

Exercise 20

1. Determine the RMS value of the periodic voltage defined as

$$v = 5.04 - 0.679 \cos \omega t - 0.432 \cos 2\omega t - 0.005 \cos 3\omega t + \ldots$$
$$+ 3.852 \sin \omega t - 1.186 \sin 2\omega t + 0.480 \sin 3\omega t + \ldots$$

2. An alternating voltage $v = f(t)$ of period T applied to a circuit produces a current $i = F(t)$ amperes of the same period.

If $$v = f(t) = 4.70 - 1.49 \cos \omega t - 0.50 \cos 2\omega t - 0.16 \cos 3\omega t + \ldots$$
$$+ 0.53 \sin \omega t - 0.52 \sin 2\omega t - 0.23 \sin 3\omega t + \ldots$$

and $$i = F(t) = 2.04 - 0.08 \cos \omega t - 0.05 \cos 2\omega t + 0.01 \cos 3\omega t + \ldots$$
$$+ 1.75 \sin \omega t - 0.36 \sin 2\omega t + 0.15 \sin 3\omega t + \ldots$$

determine the average value of the power vi over one cycle.

3. An alternating current $i = f(t)$ amperes of period T flows through a resistor of $8\ \Omega$ resistance.

$$i = f(t) = 0.52 - 8.57 \cos \omega t + 3.30 \cos 2\omega t + 0.43 \cos 3\omega t + \ldots$$
$$+ 16.33 \sin \omega t + 3.99 \sin 2\omega t + 1.13 \sin 3\omega t + \ldots$$

Determine the average value of the power p dissipated over one complete cycle, given that $p = i^2R$.

4. The Fourier series of a particular function $f(x)$ of period 2π can be expressed as

$$f(x) = \tfrac{1}{2}a_0 + \sum_{n=1}^{\infty} \{a_n \cos nx + b_n \sin nx\}$$

where
$$a_0 = -\frac{\pi}{2}$$
$$a_n = \frac{1}{\pi}\left\{\frac{\cos n\pi - 1}{n^2}\right\}$$
$$b_n = \frac{1}{n}(1 - 2\cos n\pi)$$

Determine the amplitudes of the first 10 harmonics and draw the spectrum of the function.

5. A periodic function $f(t)$ of period 5 ms is defined by

$$f(t) = 1 \qquad -2 < t < 0$$
$$f(t) = 1 - \frac{t}{3} \qquad 0 < t < 3$$
$$f(t) = f(t+5)$$

(a) Express the function as a Fourier series and show that

$$a_0 = 1.4$$
$$a_n = \frac{1}{n\pi}\sin\frac{4n\pi}{5} + \frac{5}{6n^2\pi^2}\left\{1 - \cos\frac{6n\pi}{5}\right\}$$
$$b_n = \frac{1}{n\pi}\cos\frac{4n\pi}{5} - \frac{5}{6n^2\pi^2}\sin\frac{6n\pi}{5}$$

(b) Calculate the amplitudes of the first 10 harmonics.

(c) Draw the spectrum of the waveform.

7.8 REVISION SUMMARY

1. Half-wave rectifier output

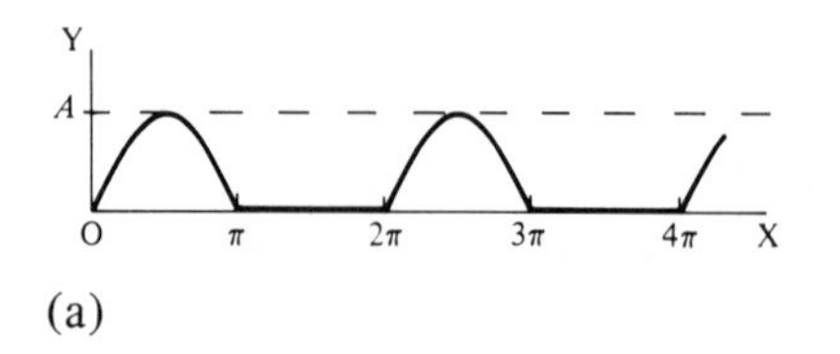

(a)

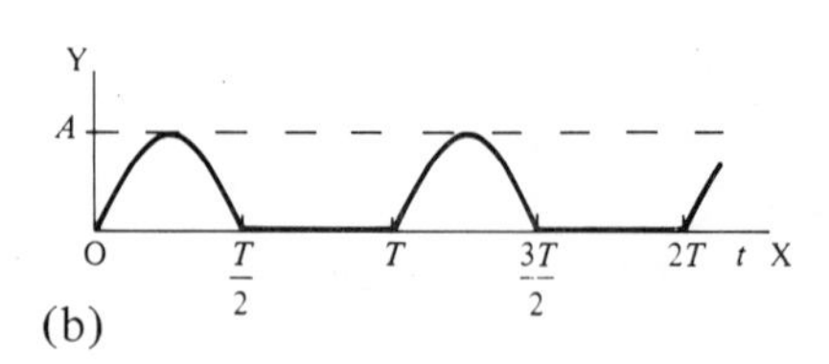

(b)

(a) *Period 2π*

$$i = \frac{A}{\pi}\left\{1 + \frac{\pi}{2}\sin x - 2\left(\frac{1}{1.3}\cos 2x + \frac{1}{3.5}\cos 4x + \ldots\right)\right\}$$

(b) *Period T*

$$i = \frac{A}{\pi}\left\{1 + \frac{\pi}{2}\sin \omega t - 2\left(\frac{1}{1.3}\cos 2\omega t + \frac{1}{3.5}\cos 4\omega t + \ldots\right)\right\}$$

where $\omega = \dfrac{2\pi}{T}$.

2. Full-wave rectifier output

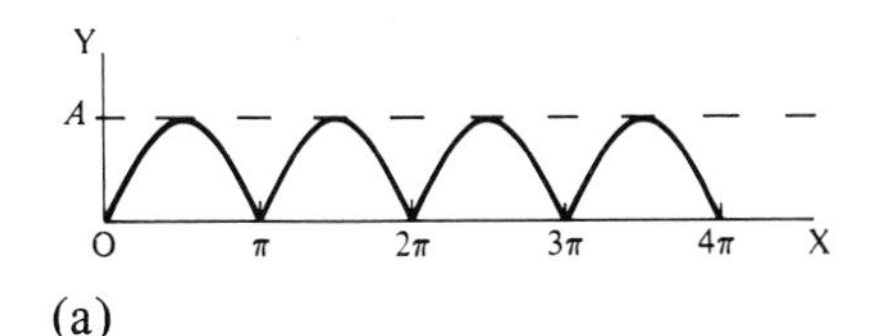

(a) (b)

(a) *Period 2π*

$$i = \frac{2A}{\pi}\left\{1 - 2\left(\frac{1}{1.3}\cos 2x + \frac{1}{3.5}\cos 4x + \frac{1}{5.7}\cos 6x + \ldots\right)\right\}$$

(b) *Period T*

$$i = \frac{2A}{\pi}\left\{1 - 2\left(\frac{1}{1.3}\cos 2\omega t + \frac{1}{3.5}\cos 4\omega t + \frac{1}{5.7}\cos 6\omega t + \ldots\right)\right\}$$

where $\omega = \dfrac{2\pi}{T}$.

3. Parseval's theorem

(a) *Period 2L*

$$f(x) = \tfrac{1}{2}a_0 + \sum_{n=1}^{\infty}\left\{a_n \cos\frac{n\pi x}{L} + b_n \sin\frac{n\pi x}{L}\right\}$$

$$\frac{1}{L}\int_{-L}^{L}\{f(x)\}^2\,\mathrm{d}x = \frac{1}{2}a_0^2 + \sum_{n=1}^{\infty}\{a_n^2 + b_n^2\}$$

(b) *Period T*

$$\frac{2}{T}\int_{0}^{T}\{f(t)\}^2\,\mathrm{d}t = \frac{1}{2}a_0^2 + \sum_{n=1}^{\infty}\{a_n^2 + b_n^2\}$$

4. RMS value

$$(\text{RMS})^2 = \tfrac{1}{4}a_0^2 + \frac{1}{2}\sum_{n=1}^{\infty}\{a_n^2 + b_n^2\}$$

When $a_0 = 0$, i.e. there is no d.c. component, and $I_n = \sqrt{a_n^2 + b_n^2}$ then

$$(\text{RMS})^2 = \frac{1}{2}\sum_{n=1}^{\infty} I_n^2 \qquad \therefore \text{ RMS} = \sqrt{\tfrac{1}{2}(I_1^2 + I_2^2 + I_3^2 + \ldots)}$$

5. Multiplication theorem

Period 2L

$$\frac{1}{L}\int_{-L}^{L} F(x)\, f(x)\, \mathrm{d}x = \tfrac{1}{2}A_0 a_0 + \sum_{n=1}^{\infty}\{A_n a_n + B_n b_n\}$$

6. Average power in a circuit over one cycle

$$p = vi = \frac{v^2}{R} = Ri^2$$

If
$$v = \tfrac{1}{2}a_0 + \sum_{n=1}^{\infty}\{a_n \cos n\omega t + b_n \sin n\omega t\}$$

and
$$i = \tfrac{1}{2}A_0 + \sum_{n=1}^{\infty}\{A_n \cos n\omega t + B_n \sin n\omega t\}$$

then
$$\text{Average value of } p = \tfrac{1}{4}A_0 a_0 + \frac{1}{2}\sum_{n=1}^{\infty}\{A_n a_n + B_n b_n\}$$

7. Spectrum of a waveform

Display of amplitudes of harmonics against a measure of frequency gives a set of discrete ordinates.

Chapter 8

FURTHER TECHNIQUES

8.1 INTEGRATION AND DIFFERENTIATION OF A FOURIER SERIES

8.1.1 Integration

A Fourier series can always be integrated term by term to obtain a further series.

For example, the function $f(x) = x$ defined in the range $-\pi < x < \pi$ with $f(x) = f(x + 2\pi)$ can be represented by the sine series

$$f(x) = x = 2\left\{\sin x - \frac{\sin 2x}{2} + \frac{\sin 3x}{3} - \frac{\sin 4x}{4} + \ldots\right\}$$

Integrating term by term, we have

$$\frac{x^2}{2} = 2\left\{-\cos x + \frac{\cos 2x}{2^2} - \frac{\cos 3x}{3^2} + \frac{\cos 4x}{4^2} + \ldots\right\} + C_1$$

$$\therefore\ x^2 = C - 4\left\{\cos x - \frac{\cos 2x}{2^2} + \frac{\cos 3x}{3^2} - \frac{\cos 4x}{4^2} + \ldots\right\}$$

The constant term C is the mean value of x^2 over the range $-\pi$ to π.

$$\therefore\ C = \frac{1}{2\pi}\int_{-\pi}^{\pi} x^2\,dx = \frac{1}{2\pi}\left[\frac{x^3}{3}\right]_{-\pi}^{\pi} = \frac{1}{6\pi}\{\pi^3 + \pi^3\} = \frac{\pi^2}{3}$$

$$\therefore\ x^2 = \frac{\pi^2}{3} - 4\left\{\cos x - \frac{\cos 2x}{2^2} + \frac{\cos 3x}{3^2} - \frac{\cos 4x}{4^2} + \ldots\right\}$$

Similarly, in general, if the function $f(x) = x$ is defined in the range $(-L, L)$ and $f(x) = f(x + 2L)$,

$$f(x) = x = \frac{2L}{\pi}\left\{\sin\frac{\pi x}{L} - \frac{1}{2}\sin\frac{2\pi x}{L} + \frac{1}{3}\sin\frac{3\pi x}{L} - \ldots\right\}$$

Then, working as before,

$$x^2 = \frac{L^2}{3} - \frac{4L^2}{\pi^2}\left\{\cos\frac{\pi x}{L} - \frac{1}{2^2}\cos\frac{2\pi x}{L} + \frac{1}{3^2}\cos\frac{3\pi x}{L} + \ldots\right\}$$

8.1.2 Differentiation

A Fourier series for $f(x)$, defined in the range $(-L, L)$ with period $2L$, can be differentiated term by term only when certain conditions are satisfied. These are that

(a) $f(x)$ is continuous in the range $(-L, L)$

(b) $f(x)$ is periodic, i.e. $f(x) = f(x + 2L)$

(c) $f'(x)$ satisfies the Dirichlet conditions, i.e. that it
  (i) is defined in the range $(-L, L)$
  (ii) is single-valued
  (iii) is piecewise continuous
  (iv) contains, at most, a finite number of finite discontinuities.

The majority of functions that occur in practical situations will be found to satisfy these requirements and can thus be differentiated term by term. Some caution, however, is necessary.

Example 1

Consider the function shown.

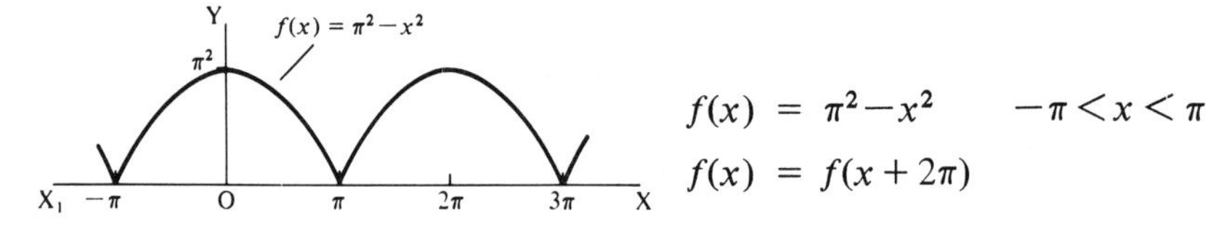

$$f(x) = \pi^2 - x^2 \qquad -\pi < x < \pi$$

$$f(x) = f(x + 2\pi)$$

$$f(x) = \pi^2 - x^2 = \frac{2\pi^2}{3} + 4\left\{\cos x - \frac{1}{2^2}\cos 2x + \frac{1}{3^2}\cos 3x - \ldots\right\}$$

In this case, $f'(x) = -2x$ and the conditions stated above are satisfied. So, differentiating term by term gives

$$-2x = 4\left\{-\sin x + \frac{2\sin 2x}{2^2} - \frac{3\sin 3x}{3^2} + \frac{4\sin 4x}{4^2} - \ldots\right\}$$

$$\therefore\ x = 2\left\{\sin x - \frac{1}{2}\sin 2x + \frac{1}{3}\sin 3x - \frac{1}{4}\sin 4x + \ldots\right\}$$

Example 2

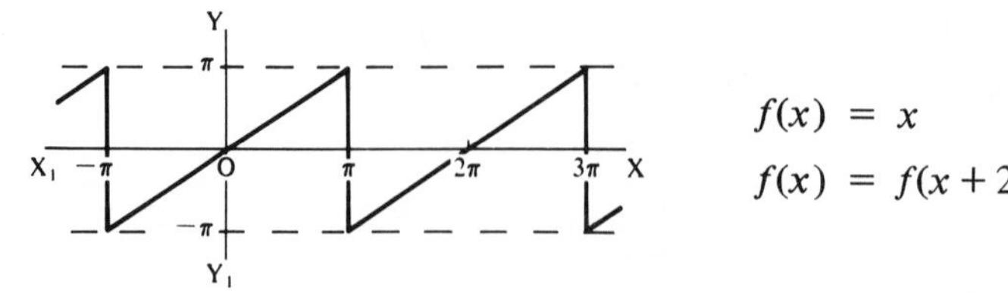

$$f(x) = x \qquad -\pi < x < \pi$$

$$f(x) = f(x + 2\pi)$$

Using the last result,

$$f(x) = x = 2\left\{\sin x - \frac{1}{2}\sin 2x + \frac{1}{3}\sin 3x - \frac{1}{4}\sin 4x + \ldots\right\}$$

and differentiating term by term, we obtain

$$1 = 2\{\cos x - \cos 2x + \cos 3x - \cos 4x + \ldots\}$$

In this case, the result obtained cannot be valid, since the series on the right hand side does not converge for any value of x.

8.2 GIBBS' PHENOMENON

The Fourier representation of a periodic function $f(x)$ defined over the interval $(-L, L)$ with $f(x) = f(x+2L)$ is

$$f(x) = \tfrac{1}{2}a_0 + \sum_{n=1}^{\infty}\left\{a_n \cos\frac{n\pi x}{L} + b_n \sin\frac{n\pi x}{L}\right\}$$

This series represents the function $f(x)$ only if an infinite number of terms are taken. When the series is restricted to a finite number (r) of terms, the partial sum $S_r(x)$ can only be an approximation to the original function $f(x)$, i.e.

$$f(x) \approx S_r(x) = \tfrac{1}{2}a_0 + \sum_{n=1}^{r}\left\{a_n \cos\frac{n\pi x}{L} + b_n \sin\frac{n\pi x}{L}\right\}$$

In practical situations, consideration of the first few harmonics will normally provide a satisfactory approximation, but the effect of such partial sums near a point of discontinuity of $f(x)$ is worthy of note.

Consider the function shown.

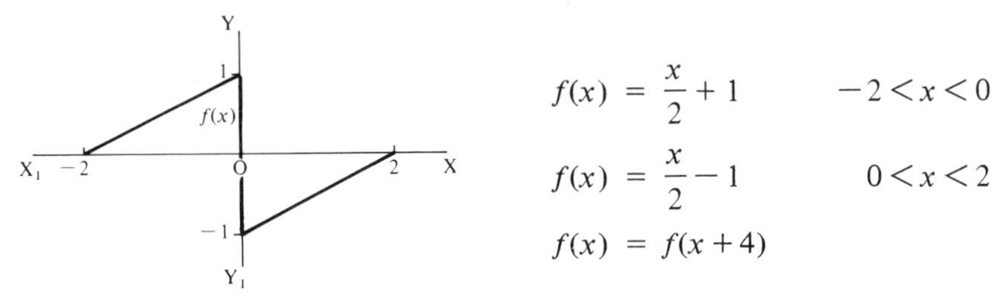

$$f(x) = \frac{x}{2} + 1 \qquad -2 < x < 0$$

$$f(x) = \frac{x}{2} - 1 \qquad 0 < x < 2$$

$$f(x) = f(x+4)$$

The function is an odd function and can therefore be expressed as a sine series.

$$a_0 = 0; \qquad a_n = 0$$

$$f(x) = \sum_{n=1}^{\infty} b_n \sin\frac{n\pi x}{2}$$

where $$b_n = \frac{1}{2}\int_{-2}^{2} f(x)\sin\frac{n\pi x}{2}\,dx$$

$$= \int_0^2 \left(\frac{x}{2} - 1\right)\sin\frac{n\pi x}{2}\,dx$$

which gives $$b_n = -\frac{2}{n\pi}$$

$$\therefore\ f(x) = -\frac{2}{\pi}\sum_{n=1}^{\infty}\frac{1}{n}\sin\frac{n\pi x}{2}$$

An approximation to the function is therefore given by the partial sum of r terms, where

$$S_r(x) = -\frac{2}{\pi}\sum_{n=1}^{r}\frac{1}{n}\sin\frac{n\pi x}{2}$$

The results obtained for $r = 1, 3, 5, 15$ are shown below.

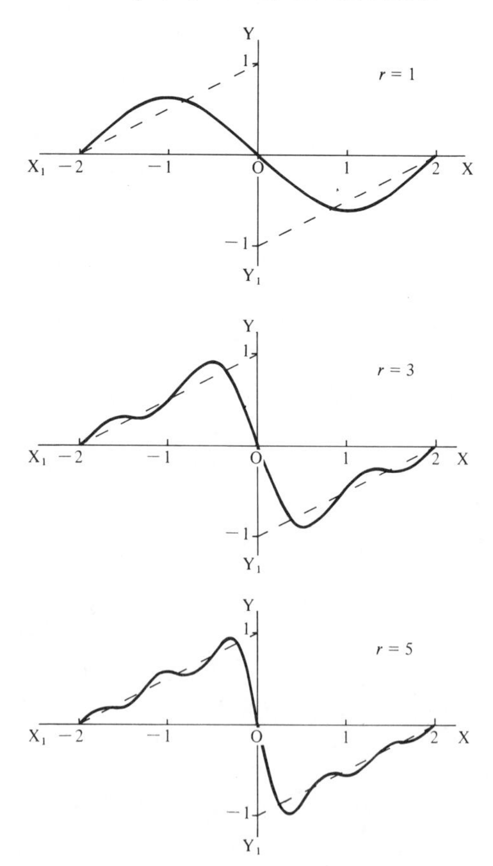

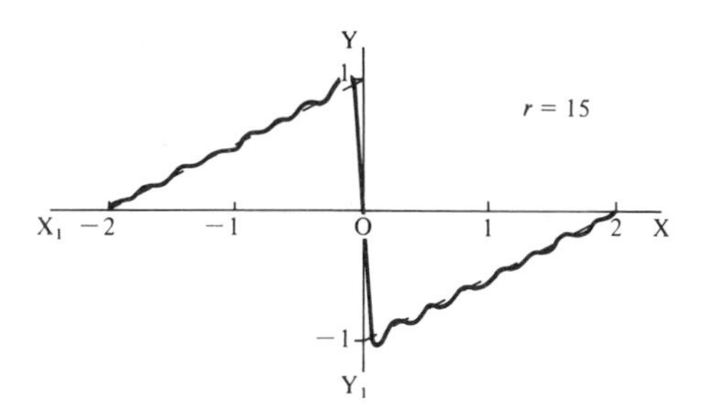

In each case, the graph of $S_r(x)$ oscillates about the line $f(x)$ with decreasing amplitudes as r is increased, progressively approximating more closely to the function it purports to represent.

In every case, however, as the point of discontinuity is approached, the graph of $S_r(x)$ overshoots both immediately before and after the discontinuity. As $r \to \infty$, the point at which the maximum value of $S_r(x)$ occurs moves closer to the point of discontinuity, but does not disappear. In fact, as $r \to \infty$, the graph tends ever nearer to the form shown below.

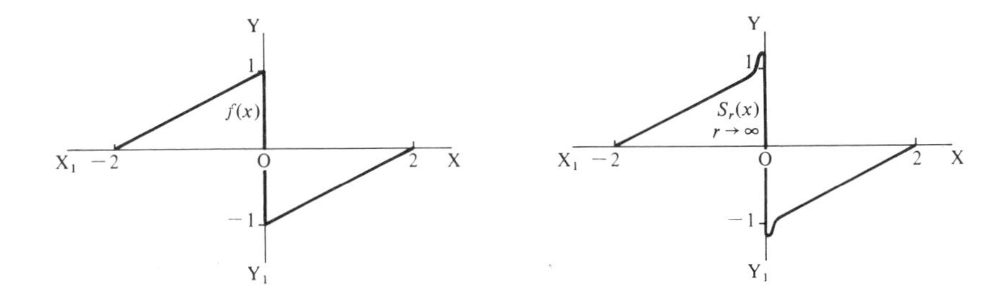

It can be proved that this overshoot will always occur at a point of discontinuity.

The Fourier series of a square waveform also shows the same characteristic

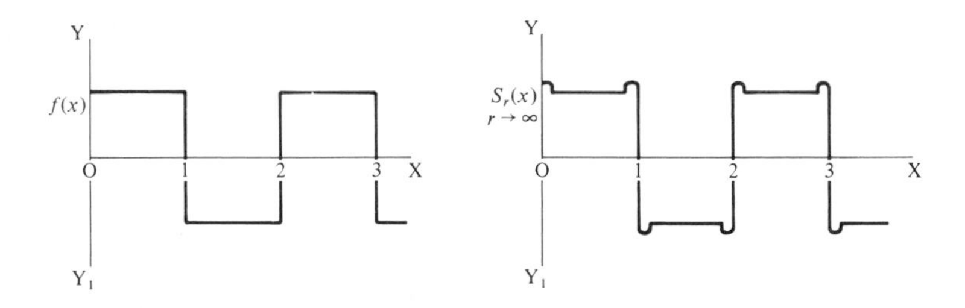

In practice, the effect of the degree of overshoot can normally be neglected, but, where special attention is being paid to the behaviour of $f(x)$ at a point of discontinuity, the local shortcoming in the Fourier representation should be appreciated.

8.3 FOURIER COEFFICIENTS FROM JUMPS AT DISCONTINUITIES

8.3.1 Positive and negative jumps

If a periodic function can be represented by arcs of polynomials, the Fourier coefficients of the series representing the function can be determined in terms of the 'jumps' of the function and its derivatives at discontinuities.

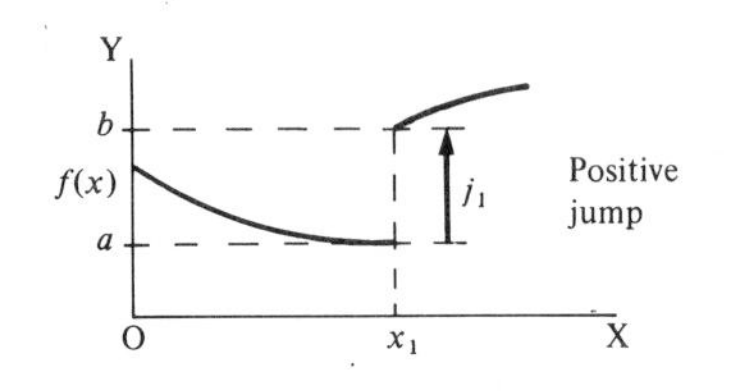

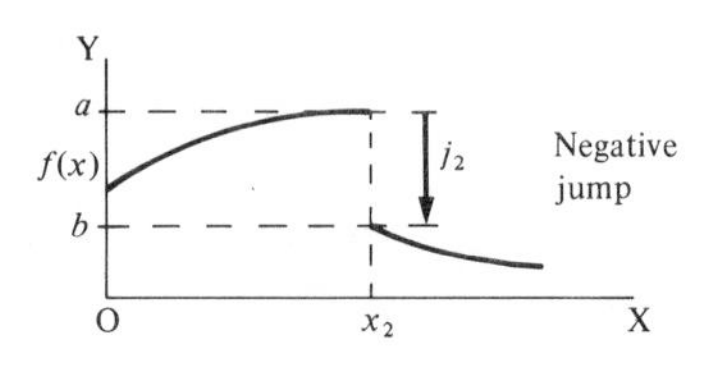

At x_1 $\qquad f(x_1-0) = a; \qquad f(x_1+0) = b$

The jump j_1 at $x = x_1$ is therefore

$$j_1 = f(x_1+0)-f(x_1-0) = b-a$$

Since in case 1, $\quad b > a,\quad$ then $\quad$ j_1 is positive

Similarly, at x_2 $\quad f(x_2-0) = a; \qquad f(x_2+0) = b$

$$j_2 = f(x_2+0)-f(x_2-0) = b-a$$

But in case 2, $\quad b < a,\quad$ then $\quad$ j_2 is negative

A function may have a finite number of finite jumps in the interval of one period.

For example

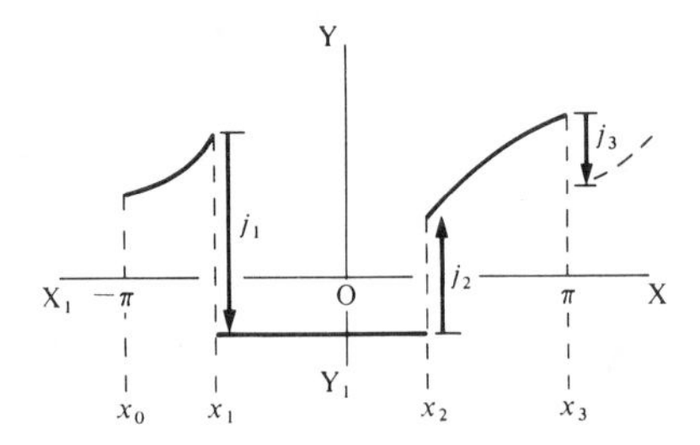

$f(x)$ is defined over $-\pi < x < \pi$ as shown and $f(x) = f(x+2\pi)$.

Jumps occur at x_1, x_2 and x_3.

j_1 is negative

j_2 is positive

j_3 is negative

Note: The initial jump at x_0 is not included, as this is regarded as the final jump in the previous cycle.

Example

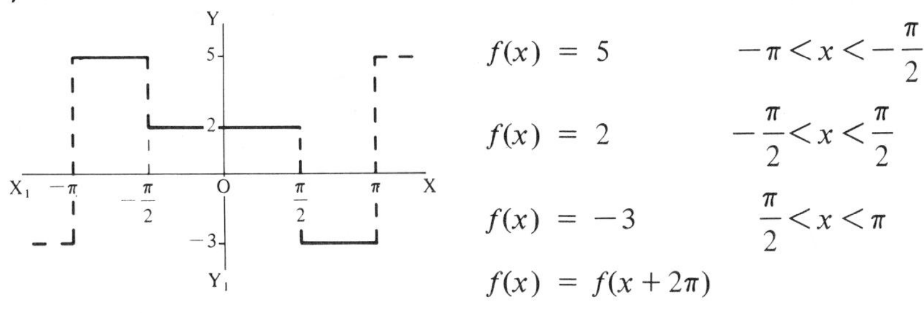

$$f(x) = 5 \qquad -\pi < x < -\frac{\pi}{2}$$

$$f(x) = 2 \qquad -\frac{\pi}{2} < x < \frac{\pi}{2}$$

$$f(x) = -3 \qquad \frac{\pi}{2} < x < \pi$$

$$f(x) = f(x+2\pi)$$

Jumps occur at $\quad x_1 = -\dfrac{\pi}{2}; \qquad x_2 = \dfrac{\pi}{2}; \qquad x_3 = \pi$

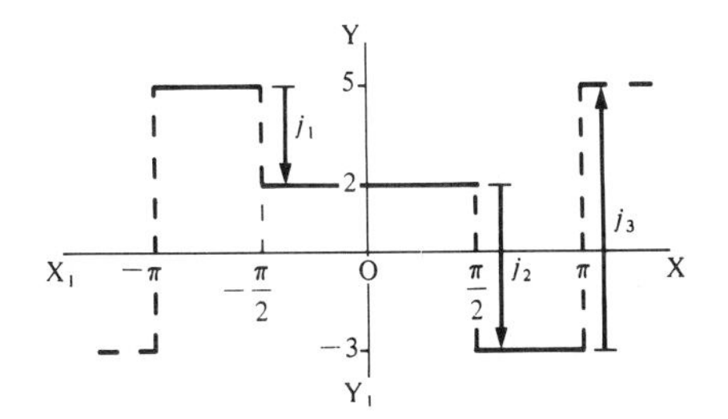

$$\therefore\ j_1 = -3$$
$$j_2 = -5$$
$$j_3 = 8$$

Exercise 21

Determine the values of the jumps in one cycle of the following periodic functions, each defined over an interval of 2π with $f(x) = f(x+2\pi)$.

1. $f(x) = -2 \qquad -\pi < x < 0$

   $f(x) = 3x \qquad 0 < x < \pi$

2. $f(x) = 0 \qquad -\pi < x < 0$

   $f(x) = x^2 \qquad 0 < x < \pi$

3. $f(x) = x(2\pi - x) \qquad 0 < x < \pi$

   $f(x) = 4 \qquad \pi < x < 2\pi$

4. $f(x) = 5 \qquad 0 < x < \dfrac{\pi}{2}$

   $f(x) = 2 - \dfrac{x^3}{10} \qquad \dfrac{\pi}{2} < x < \dfrac{3\pi}{2}$

   $f(x) = 5 \qquad \dfrac{3\pi}{2} < x < 2\pi$

5. $f(x) = \pi + x \qquad -\pi < x < 0$

   $f(x) = \dfrac{\pi}{2} - x \qquad 0 < x < \pi$

6. $f(x) = -4 \qquad -\pi < x < -\frac{\pi}{2}$

$f(x) = 4 \qquad -\frac{\pi}{2} < x < \frac{\pi}{2}$

$f(x) = -4 \qquad \frac{\pi}{2} < x < \pi$

7. $f(x) = 7 - x^2 \qquad -\pi < x < 0$

$f(x) = x^2 - 7 \qquad 0 < x < \pi$

8. $f(x) = \frac{4x}{\pi} + 4 \qquad -\pi < x < -\frac{\pi}{2}$

$f(x) = -2 \qquad -\frac{\pi}{2} < x < \frac{\pi}{2}$

$f(x) = \frac{4x}{\pi} - 2 \qquad \frac{\pi}{2} < x < \pi$

9. $f(x) = 2x \qquad 0 < x < \frac{\pi}{2}$

$f(x) = \pi \qquad \frac{\pi}{2} < x < \pi$

$f(x) = 2(\pi - x) \qquad \pi < x < \frac{3\pi}{2}$

$f(x) = -\pi \qquad \frac{3\pi}{2} < x < 2\pi$

10. $f(x) = 2 + x^2 \qquad 0 < x < \frac{\pi}{3}$

$f(x) = \left(\frac{\pi}{3} - x\right)\left(\frac{2\pi}{3} - x\right) \qquad \frac{\pi}{3} < x < \frac{2\pi}{3}$

$f(x) = 0 \qquad \frac{2\pi}{3} < x < 2\pi$

8.3.2 Notation

Jumps may also occur in the derivatives of the function. These we can denote by primes (or dashes) attached to the symbol j, thus

j_1' for the jump in the first derivative at $x = x_1$
j_1'' for the jump in the second derivative at $x = x_1$
j_1''' for the jump in the third derivative at $x = x_1$

etc.

So j_3'' denotes the jump in the second derivative at $x = x_3$

8.3.3 Determination of Fourier coefficients

For a function $f(x)$ defined in the interval $-\pi < x < \pi$ with $f(x) = f(x+2\pi)$,

$$f(x) = \tfrac{1}{2}a_0 + \sum_{n=1}^{\infty}\{a_n \cos nx + b_n \sin nx\}$$

where

$$a_0 = \frac{1}{\pi}\int_{-\pi}^{\pi} f(x)\,\mathrm{d}x \tag{8.1}$$

$$a_n = \frac{1}{\pi}\int_{-\pi}^{\pi} f(x)\cos nx\,\mathrm{d}x \tag{8.2}$$

$$b_n = \frac{1}{\pi}\int_{-\pi}^{\pi} f(x)\sin nx\,\mathrm{d}x \tag{8.3}$$

From (8.2)

$$\pi a_n = \int_{-\pi}^{\pi} f(x)\cos nx\,\mathrm{d}x$$

$$= \left[f(x)\frac{\sin nx}{n}\right]_{-\pi}^{\pi} - \frac{1}{n}\int_{-\pi}^{\pi}\sin nx\, f'(x)\,\mathrm{d}x \tag{8.4}$$

Let $f(x)$ contains r finite discontinuities in the range $-\pi < x < \pi$ at $x_1, x_2, x_3, \ldots, x_r$.

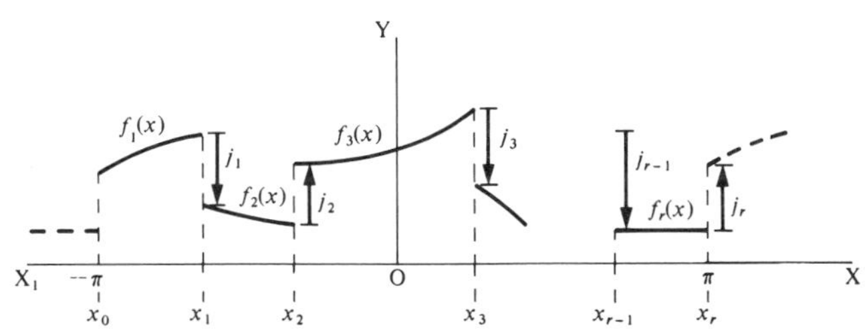

Then

$$\left[f(x)\frac{\sin nx}{n}\right]_{-\pi}^{\pi} = \left[f_1(x)\frac{\sin nx}{n}\right]_{x_0}^{x_1} + \left[f_2(x)\frac{\sin nx}{n}\right]_{x_1}^{x_2} + \left[f_3(x)\frac{\sin nx}{n}\right]_{x_2}^{x_3}$$

$$+ \ldots + \left[f_r(x)\frac{\sin nx}{n}\right]_{x_{r-1}}^{x_r} \tag{8.5}$$

Considering the first term on the right hand side of (8.5),

$$\frac{1}{n}\left[f_1(x)\sin nx\right]_{x_0}^{x_1} = \frac{1}{n}\{f_1(x_1)\sin nx_1 - f_1(x_0)\sin nx_0\}$$

Since discontinuities occur at $x = x_0, x_1, x_2, \ldots, x_r$, this must be written as

$$\frac{1}{n}\{f_1(x_1 - 0)\sin nx_1 - f_1(x_0 + 0)\sin nx_0\}$$

The right hand side of (8.5) then becomes

$$\begin{aligned}\text{RHS} = \frac{1}{n}\{ & f_1(x_1-0)\sin nx_1 - f_1(x_0+0)\sin nx_0 \\ & + f_2(x_2-0)\sin nx_2 - f_2(x_1+0)\sin nx_1 \\ & + f_3(x_3-0)\sin nx_3 - f_3(x_2+0)\sin nx_2 \\ & \cdot \\ & \cdot \\ & \cdot \\ & + f_r(x_r-0)\sin nx_r - f_r(x_{r-1}+0)\sin nx_{r-1}\}\end{aligned}$$

This can be re-written in the form

$$\begin{aligned} n \times \text{RHS} = & -f_1(x_0+0)\sin nx_0 \\ & -\{f_2(x_1+0)-f_1(x_1-0)\}\sin nx_1 \\ & -\{f_3(x_2+0)-f_2(x_2-0)\}\sin nx_2 \\ & \cdot \\ & \cdot \\ & \cdot \\ & -\{f_r(x_{r-1}+0)-f_{r-1}(x_{r-1}-0)\}\sin nx_{r-1} \\ & + f_r(x_r-0)\sin nx_r\end{aligned}$$

Now $\sin nx_0 = \sin nx_r$. Therefore the first and last terms give

$$\begin{aligned} & -f_1(x_0+0)\sin nx_0 + f_r(x_r-0)\sin nx_r \\ & = -\{f_1(x_0+0)-f_r(x_r-0)\}\sin nx_r\end{aligned}$$

Note also that $\{f_r(x_{r-1}+0)-f_{r-1}(x_{r-1}-0)\}$ is the jump at $x = x_{r-1}$.

Then (8.4) finally becomes

$$n\pi a_n = -j_1 \sin nx_1 - j_2 \sin nx_2 - j_3 \sin nx_3 - \ldots - j_r \sin nx_r$$
$$-\int_{-\pi}^{\pi} f'(x) \sin nx \, dx$$

$$\therefore \; \pi a_n = -\frac{1}{n}\sum_{r=1}^{r} j_r \sin nx_r - \frac{1}{n}\int_{-\pi}^{\pi} f'(x)\sin nx \, dx$$

Discontinuities in $f'(x)$ may well also occur at $x_0, x_1, x_2, \ldots, x_r$ and $\int_{-\pi}^{\pi} f'(x) \sin nx \, dx$ can be treated in the same way to give

$$\int_{-\pi}^{\pi} f'(x)\sin nx \, dx = \frac{1}{n}\sum_{r=1}^{r} j_r' \cos nx_r + \frac{1}{n}\int_{-\pi}^{\pi} f''(x)\cos nx \, dx$$

Continuing the process for the right hand integral involves expressions of successively higher derivatives of $f(x)$ and since $f(x)$ has been defined as being represented by arcs of polynomials of x, eventually the derivatives will reach zero.

Therefore, collecting results, we obtain

$$a_n = \frac{1}{n\pi}\left\{-\sum_{r=1}^{r} j_r \sin nx_r - \frac{1}{n}\sum_{r=1}^{r} j_r' \cos nx_r + \frac{1}{n^2}\sum_{r=1}^{r} j_r'' \sin nx_r + \frac{1}{n^3}\sum_{r=1}^{r} j_r''' \cos nx_r \quad - \quad - \quad + \quad + \quad \ldots\right\}, \qquad n = 1,2,3,\ldots \tag{8.6}$$

The working for b_n follows the same steps, finally giving

$$b_n = \frac{1}{n\pi}\left\{\sum_{r=1}^{r} j_r \cos nx_r - \frac{1}{n}\sum_{r=1}^{r} j_r' \sin nx_r - \frac{1}{n^2}\sum_{r=1}^{r} j_r'' \cos nx_r + \frac{1}{n^3}\sum_{r=1}^{r} j_r''' \sin nx_r \quad + \quad - \quad - \quad + \quad + \quad \ldots\right\} \qquad n = 1,2,3,\ldots \tag{8.7}$$

Note that a_0 is obtained by direct integration in the normal way.

$$a_0 = \frac{1}{\pi}\int_{-\pi}^{\pi} f(x)\,dx$$

Example 1

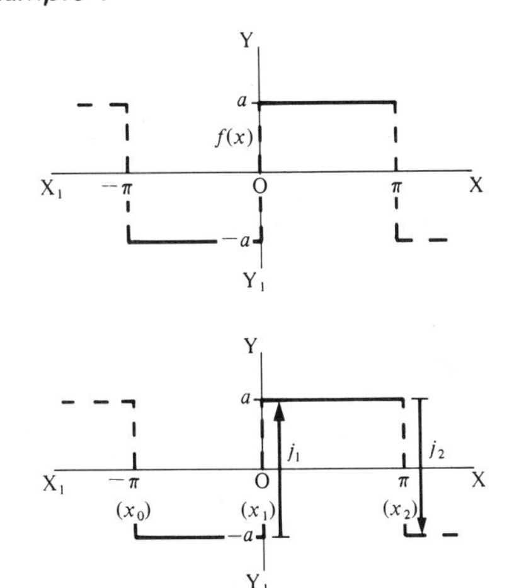

$$f(x) = -a \qquad -\pi < x < 0$$

$$f(x) = a \qquad 0 < x < \pi$$

$$f(x) = f(x + 2\pi)$$

Jumps occur as follows

$$j_1 = 2a \quad \text{at} \quad x = x_1 = 0$$

$$j_2 = -2a \quad \text{at} \quad x = x_2 = \pi$$

All derivatives are zero.

x	$x_1 = 0$	$x_2 = \pi$
$f(x)$	$j_1 = 2a$	$j_2 = -2a$
$f'(x)$	$j_1' = 0$	$j_2' = 0$

The function is odd. Therefore the series contains only sine terms.

$$b_n = \frac{1}{n\pi}\left\{\sum_{r=1}^{r} j_r \cos nx_r\right\}$$

All other terms are zero since the derivatives are zero.

$$= \frac{1}{n\pi}\{j_1 \cos nx_1 + j_2 \cos nx_2\}$$

$$= \frac{1}{n\pi}\{2a \cos 0 - 2a \cos n\pi\}$$

$$= \frac{1}{n\pi}\{2a - 2a(-1)^n\} = \frac{2a}{n\pi}\{1-(-1)^n\}$$

$$\therefore\ f(x) = \sum_{n=1}^{\infty} b_n \sin nx = \sum_{n=1}^{\infty} \frac{2a}{n\pi}\{1-(-1)^n\}\sin nx$$

i.e. $$f(x) = \frac{4a}{\pi}\left\{\sin x + \frac{1}{3}\sin 3x + \frac{1}{5}\sin 5x + \ldots\right\}$$

Example 2

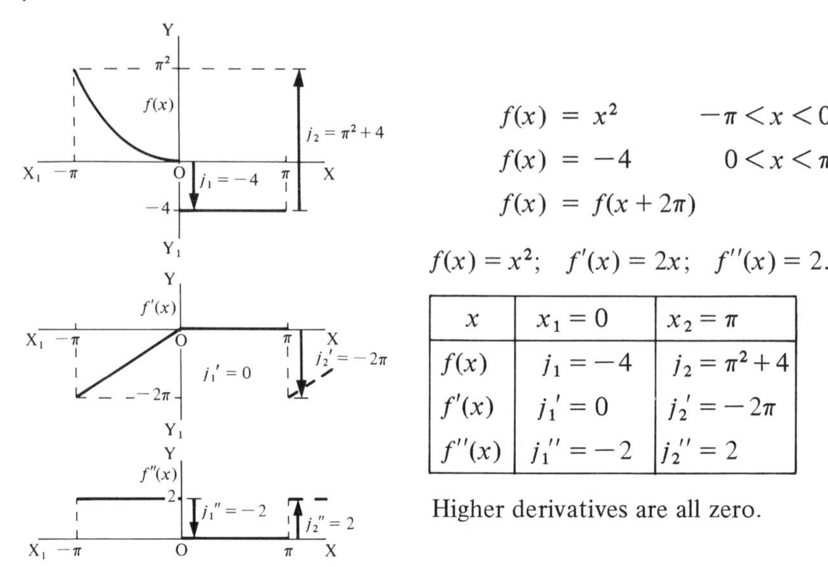

$$f(x) = x^2 \qquad -\pi < x < 0$$
$$f(x) = -4 \qquad 0 < x < \pi$$
$$f(x) = f(x + 2\pi)$$

$f(x) = x^2;\quad f'(x) = 2x;\quad f''(x) = 2.$

x	$x_1 = 0$	$x_2 = \pi$
$f(x)$	$j_1 = -4$	$j_2 = \pi^2 + 4$
$f'(x)$	$j_1' = 0$	$j_2' = -2\pi$
$f''(x)$	$j_1'' = -2$	$j_2'' = 2$

Higher derivatives are all zero.

$$a_0 = \frac{1}{\pi}\int_{-\pi}^{\pi} f(x)\,dx = \frac{1}{\pi}\left\{\int_{-\pi}^{0} x^2\,dx + \int_{0}^{\pi}(-4)\,dx\right\}$$

$$= \frac{1}{\pi}\left\{\left[\frac{x^3}{3}\right]_{-\pi}^{0} - 4\Big[x\Big]_0^{\pi}\right\} = \frac{1}{\pi}\left\{\frac{\pi^3}{3} - 4\pi\right\} \qquad \therefore\ a_0 = \frac{\pi^2}{3} - 4$$

$$a_n = \frac{1}{n\pi}\left\{-\sum_{r=1} j_r \sin nx_r - \frac{1}{n}\sum_{r=1} j_r' \cos nx_r + \frac{1}{n^2}\sum_{r=1}^{r} j_r'' \sin nx_r\right\}$$

$$= \frac{1}{n\pi}\left\{-j_1 \sin nx_1 - j_2 \sin nx_2 - \frac{1}{n} j_1' \cos nx_1 - \frac{1}{n} j_2' \cos nx_2 + \frac{1}{n^2} j_1'' \sin nx_1 + \frac{1}{n^2} j_2'' \sin nx_2\right\}$$

$$= \frac{1}{n\pi}\left\{4 \sin 0 - (\pi^2+4)\sin n\pi - \frac{1}{n}(0)\cos 0 + \frac{1}{n}2\pi \cos n\pi + \frac{1}{n^2}(-2)\sin 0 + \frac{2}{n^2}\sin n\pi\right\}$$

$$= \left(\frac{1}{n\pi}\right)\left(\frac{2\pi}{n}\right)\cos n\pi = \frac{2}{n^2}\cos n\pi = \frac{2}{n^2}(-1)^n \qquad \therefore\ a_n = \frac{2}{n^2}(-1)^n$$

$$b_n = \frac{1}{n\pi}\left\{\sum_{r=1}^{r} j_r \cos nx_r - \frac{1}{n}\sum_{r=1}^{r} j_r' \sin nx_r - \frac{1}{n^2}\sum_{r=1}^{r} j_r'' \cos nx_r\right\}$$

$$= \frac{1}{n\pi}\left\{j_1 \cos nx_1 + j_2 \cos nx_2 - \frac{1}{n} j_1' \sin nx_1 - \frac{1}{n} j_2' \sin nx_2 - \frac{1}{n^2} j_1'' \cos nx_1 - \frac{1}{n^2} j_2'' \cos nx_2\right\}$$

$$= \frac{1}{n\pi}\left\{-4\cos 0 + (\pi^2+4)\cos n\pi - \frac{1}{n}(0)\sin 0 - \frac{1}{n}(-2\pi)\sin n\pi - \frac{1}{n^2}(-2)\cos 0 - \frac{1}{n^2}2\cos n\pi\right\}$$

$$= \frac{1}{n\pi}\left\{\left(-4 + \frac{2}{n^2}\right)\cos 0 + \left(\pi^2 + 4 - \frac{2}{n^2}\right)\cos n\pi\right\}$$

$$= \frac{1}{n\pi}\left\{\frac{2}{n^2}(1-\cos n\pi) - 4 + (\pi^2+4)\cos n\pi\right\}$$

$$\therefore\ b_n = \frac{1}{n\pi}\left\{\frac{2}{n^2}[1-(-1)^n] - 4 + (\pi^2+4)(-1)^n\right\}$$

$$\therefore\ b_1 = -\frac{\pi^2+4}{\pi}; \quad b_2 = \frac{\pi}{2}; \quad b_3 = -\frac{1}{3\pi}\left(\pi^2 + \frac{68}{9}\right); \quad b_4 = \frac{\pi}{4}$$

$$\therefore f(x) = \frac{\pi^2}{6} - 2 + \sum_{n=1}^{\infty} \frac{2}{n^2}(-1)^n \cos nx$$

$$+ \sum_{n=1}^{\infty} \frac{1}{n\pi}\left\{\frac{2}{n^2}[1-(-1)^n] - 4 + (\pi^2+4)(-1)^n\right\} \sin nx$$

Example 3

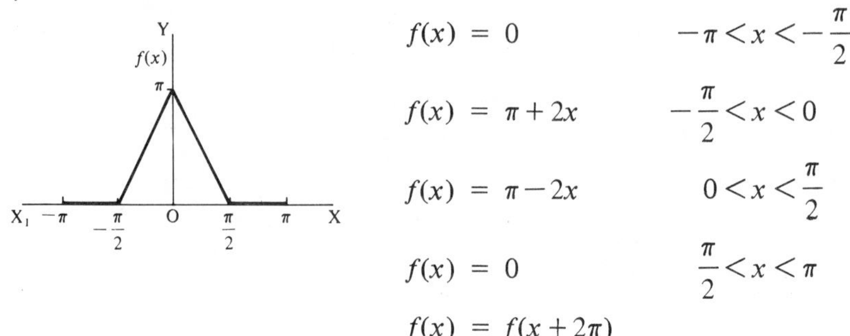

$$f(x) = 0 \qquad -\pi < x < -\frac{\pi}{2}$$

$$f(x) = \pi + 2x \qquad -\frac{\pi}{2} < x < 0$$

$$f(x) = \pi - 2x \qquad 0 < x < \frac{\pi}{2}$$

$$f(x) = 0 \qquad \frac{\pi}{2} < x < \pi$$

$$f(x) = f(x + 2\pi)$$

$f(x)$ is an even function. Hence, there are no sine terms in the series.

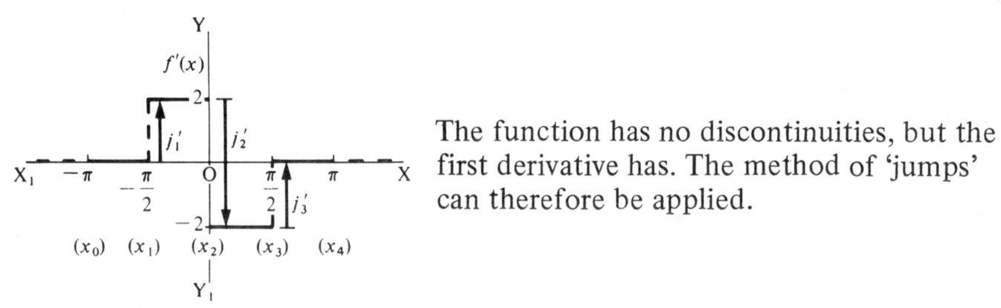

The function has no discontinuities, but the first derivative has. The method of 'jumps' can therefore be applied.

x	$x_1 = -\frac{\pi}{2}$	$x_2 = 0$	$x_3 = \frac{\pi}{2}$	$x_4 = \pi$
$f(x)$	$j_1 = 0$	$j_2 = 0$	$j_3 = 0$	$j_4 = 0$
$f'(x)$	$j_1' = 2$	$j_2' = -4$	$j_3' = 2$	$j_4' = 0$
$f''(x)$	$j_1'' = 0$	$j_2'' = 0$	$j_3'' = 0$	$j_4'' = 0$

$$a_0 = \frac{1}{\pi}\int_{-\pi}^{\pi} f(x)\,dx$$

$$\therefore \pi a_0 = \int_{-\pi/2}^{0} (\pi + 2x)\,dx + \int_{0}^{\pi/2} (\pi - 2x)\,dx$$

$$= \left[\pi x + x^2\right]_{-\pi/2}^{0} + \left[\pi x - x^2\right]_{0}^{\pi/2} \qquad \therefore a_0 = \frac{\pi}{2}$$

$$a_n = \frac{1}{n\pi}\left\{-\sum_{r=1}^{r} j_r \sin nx_r - \frac{1}{n}\sum_{r=1}^{r} j_r' \cos nx_r\right\}$$

But $j_1 = j_2 = j_3 = j_4 = 0$.

$$\therefore\ a_n = \frac{1}{n\pi}\left\{-\frac{1}{n}\sum_{r=1}^{r} j_r' \cos nx_r\right\}$$

$$= \frac{1}{n\pi}\left\{-\frac{1}{n}\left[2\cos n\left(-\frac{\pi}{2}\right) - 4\cos n(0) + 2\cos\frac{n\pi}{2} + (0)\cos n\pi\right]\right\}$$

$$= -\frac{4}{n^2\pi}\left\{\cos\frac{n\pi}{2} - 1\right\}$$

$$\therefore\ f(x) = \frac{\pi}{4} - \frac{4}{\pi}\sum_{n=1}^{\infty}\frac{1}{n^2}\left(\cos\frac{n\pi}{2} - 1\right)\cos nx$$

$$\therefore\ f(x) = \frac{\pi}{4} + \frac{4}{\pi}\left\{\cos x + \frac{1}{2}\cos 2x + \frac{1}{3^2}\cos 3x + \frac{1}{5^2}\cos 5x + \ldots\right\}$$

Exercise 22

Use the method of 'jumps' to determine the Fourier series representing the following functions. In each case, $f(x) = f(x + 2\pi)$.

1. $f(x) = x(x + 2\pi)$ $\qquad 0 < x < 2\pi$

2. $f(x) = -4$ $\qquad -\pi < x < 0$

   $f(x) = 2x$ $\qquad 0 < x < \pi$

3. $f(x) = x$ $\qquad 0 < x < \frac{\pi}{2}$

   $f(x) = \frac{\pi}{2}$ $\qquad \frac{\pi}{2} < x < \frac{3\pi}{2}$

   $f(x) = 2\pi - x$ $\qquad \frac{3\pi}{2} < x < 2\pi$

4. $f(x) = \frac{3x}{\pi} + 5$ $\qquad -\pi < x < 0$

   $f(x) = -\left(3 + \frac{2x}{\pi}\right)$ $\qquad 0 < x < \pi$

5. $f(x) = 2 \qquad -\pi < x < -\frac{\pi}{2}$

$f(x) = -\frac{4x}{\pi} \qquad -\frac{\pi}{2} < x < \frac{\pi}{2}$

$f(x) = -2 \qquad \frac{\pi}{2} < x < \pi$

6. $f(x) = x + \pi \qquad -\pi < x < 0$

$f(x) = \pi \qquad 0 < x < \frac{\pi}{2}$

$f(x) = 0 \qquad \frac{\pi}{2} < x < \pi$

7. $f(x) = x^2 \qquad 0 < x < \pi$

$f(x) = -2 \qquad \pi < x < 2\pi$

8. $f(x) = 2(x + \pi) \qquad -\pi < x < \frac{\pi}{2}$

$f(x) = 3\pi \qquad \frac{\pi}{2} < x < \pi$

9. $f(x) = 2 \qquad -\pi < x < -\frac{\pi}{2}$

$f(x) = x^3 \qquad -\frac{\pi}{2} < x < \frac{\pi}{2}$

$f(x) = -2 \qquad \frac{\pi}{2} < x < \pi$

10. $f(x) = \frac{\pi^2}{4} \qquad 0 < x < \frac{\pi}{2}$

$f(x) = (x - \pi)^2 \qquad \frac{\pi}{2} < x < \frac{3\pi}{2}$

$f(x) = 0 \qquad \frac{3\pi}{2} < x < 2\pi$

8.4 COMPLEX FORM OF FOURIER SERIES

8.4.1 Determination of Fourier coefficients

(a) *Period 2π*

$$f(x) = \tfrac{1}{2}a_0 + \sum_{n=1}^{\infty}\{a_n \cos nx + b_n \sin nx\}$$

where
$$a_0 = \frac{1}{\pi}\int_0^{2\pi} f(x)\,\mathrm{d}x$$

$$a_n = \frac{1}{\pi}\int_0^{2\pi} f(x)\cos nx\,\mathrm{d}x$$

$$b_n = \frac{1}{\pi}\int_0^{2\pi} f(x)\sin nx\,\mathrm{d}x$$

$$\therefore\ a_n + \mathrm{j}b_n = \frac{1}{\pi}\left\{\int_0^{2\pi} f(x)\cos nx\,\mathrm{d}x + \mathrm{j}\int_0^{2\pi} f(x)\sin nx\,\mathrm{d}x\right\}$$

$$= \frac{1}{\pi}\left\{\int_0^{2\pi} f(x)\,(\cos nx + \mathrm{j}\sin nx)\,\mathrm{d}x\right\}$$

But $\cos\theta + \mathrm{j}\sin\theta = \mathrm{e}^{\mathrm{j}\theta}$

$$\therefore\ \underline{a_n + \mathrm{j}b_n = \frac{1}{\pi}\int_0^{2\pi} f(x)\ \mathrm{e}^{\mathrm{j}nx}\,\mathrm{d}x} \qquad n = 1, 2, 3, \ldots$$

Evaluating the integral and equating real and imaginary parts gives expressions for a_n and b_n.

(b) *Period $2L$*

The result above becomes

$$\underline{a_n + \mathrm{j}b_n = \frac{1}{L}\int_{-L}^{L} f(x)\ \mathrm{e}^{\mathrm{j}n\pi x/L}\,\mathrm{d}x} \qquad n = 1, 2, 3, \ldots$$

(c) *Period T*

$$\omega = 2\pi f = \frac{2\pi}{T} \qquad \therefore\ T = \frac{2\pi}{\omega}$$

$$\underline{a_n + \mathrm{j}b_n = \frac{2}{T}\int_0^{T} f(t)\ \mathrm{e}^{\mathrm{j}n\omega t}\,\mathrm{d}t} \qquad \text{where}\quad \omega = \frac{2\pi}{T}$$

Example 1

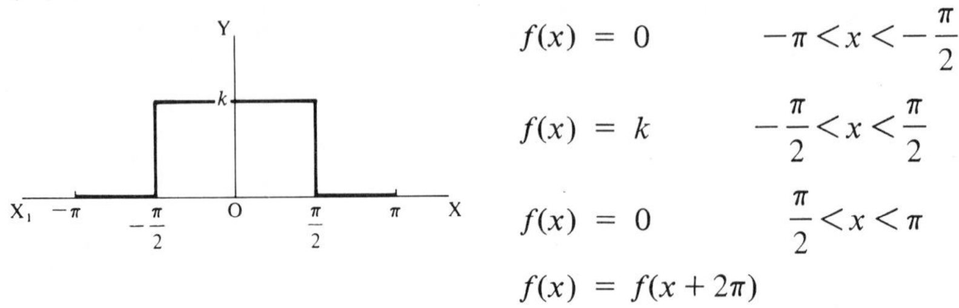

$$f(x) = 0 \qquad -\pi < x < -\frac{\pi}{2}$$

$$f(x) = k \qquad -\frac{\pi}{2} < x < \frac{\pi}{2}$$

$$f(x) = 0 \qquad \frac{\pi}{2} < x < \pi$$

$$f(x) = f(x+2\pi)$$

$$a_n + \mathrm{j}b_n = \frac{1}{\pi}\int_0^{2\pi} f(x)\,\mathrm{e}^{\mathrm{j}nx}\,\mathrm{d}x$$

$$= \frac{1}{\pi}\int_{-\pi/2}^{\pi/2} k\,\mathrm{e}^{\mathrm{j}nx}\,\mathrm{d}x$$

$$= \frac{k}{\pi}\left[\frac{\mathrm{e}^{\mathrm{j}nx}}{\mathrm{j}n}\right]_{-\pi/2}^{\pi/2} = \frac{k}{\pi \mathrm{j}n}\{\mathrm{e}^{\mathrm{j}n\pi/2} - \mathrm{e}^{-\mathrm{j}n\pi/2}\}$$

$$= \frac{k}{\pi \mathrm{j}n}\left\{\left(\cos n\frac{\pi}{2} + \mathrm{j}\sin n\frac{\pi}{2}\right) - \left(\cos n\frac{\pi}{2} - \mathrm{j}\sin n\frac{\pi}{2}\right)\right\}$$

$$= \frac{k}{\pi \mathrm{j}n}\left\{\mathrm{j}2\sin n\frac{\pi}{2}\right\} = \frac{2k}{\pi}\left\{\frac{1}{n}\sin n\frac{\pi}{2}\right\} \quad \text{i.e. entirely real}$$

$$\therefore\ \underline{a_n = \frac{2k}{\pi}\left\{\frac{1}{n}\sin n\frac{\pi}{2}\right\}} \quad \text{and} \quad \underline{b_n = 0}$$

$$a_0 = \frac{1}{\pi}\int_{-\pi}^{\pi} f(x)\,\mathrm{d}x = \frac{k}{\pi}\int_{-\pi/2}^{\pi/2}\mathrm{d}x = \frac{k}{\pi}\Big[x\Big]_{-\pi/2}^{\pi/2} = k \quad \therefore\ \underline{\tfrac{1}{2}a_0 = \frac{k}{2}}$$

$$f(x) = \frac{k}{2} + \sum_{n=1}^{\infty}\left(\frac{2k}{\pi}\right)\left(\frac{1}{n}\right)\sin n\frac{\pi}{2}\cos nx$$

$$\therefore\ \underline{f(x) = k\left\{\frac{1}{2} + \frac{2}{\pi}\sum_{n=1}^{\infty}\frac{1}{n}\sin\frac{n\pi}{2}\cos nx\right\}} \qquad n = 1, 2, 3, \ldots$$

Example 2

$$f(t) = \mathrm{e}^t \qquad 0 < t < 2$$

$$f(t) = f(t+2)$$

$$f(t) = \tfrac{1}{2}a_0 + \sum_{n=1}^{\infty}\{a_n \cos n\omega t + b_n \sin n\omega t\}$$

where $\omega = 2\pi f = \dfrac{2\pi}{T} = \pi.$

$$a_0 = \frac{2}{2}\int_0^2 f(t)\,dt = \int_0^2 e^t\,dt = \Big[e^t\Big]_0^2 = e^2 - 1$$

$$\therefore\ \tfrac{1}{2}a_0 = \frac{e^2-1}{2}$$

$$a_n + jb_n = \frac{2}{2}\int_0^2 f(t)\,e^{jn\omega t}\,dt$$

$$= \int_0^2 e^t\,e^{jn\omega t}\,dt = \int_0^2 e^{(1+jn\omega)t}\,dt$$

$$= \left[\frac{e^{(1+jn\omega)t}}{1+jn\omega}\right]_0^2 = \frac{e^{(2+j2n\omega)} - 1}{1+jn\omega}$$

But $\omega = \pi$

$$\therefore\ a_n + jb_n = \frac{1-jn\omega}{1+n^2\omega^2}\{e^2(\cos 2n\pi + j\sin 2n\pi) - 1\}$$

But $\left.\begin{aligned}\cos 2n\pi &= 1\\ \sin 2n\pi &= 0\end{aligned}\right\}\quad n = 1, 2, 3, \ldots$

$$\therefore\ a_n + jb_n = \frac{1-jn\omega}{1+n^2\omega^2}\{e^2 - 1\} = \frac{e^2-1}{1+n^2\omega^2}\{1 - jn\omega\}$$

Equating real and imaginary parts, with $\omega = \pi$

$$a_n = \frac{e^2-1}{1+n^2\pi^2} \quad \text{and} \quad b_n = -\frac{(e^2-1)n\pi}{1+n^2\pi^2}$$

$$\therefore\ f(t) = (e^2-1)\left\{\frac{1}{2} + \sum_{n=1}^{\infty}\frac{\cos n\pi t - n\pi \sin n\pi t}{1+n^2\pi^2}\right\}$$

8.4.2 Complex form of Fourier series

(a) *Period 2π*

$$f(x) = \tfrac{1}{2}a_0 + \sum_{n=1}^{\infty}\{a_n \cos nx + b_n \sin nx\} \qquad n = 1, 2, 3, \ldots$$

$$= \tfrac{1}{2}a_0 + \sum_{n=1}^{\infty}\{k_n \cos(nx - \phi_n)\}$$

where
$$a_n = k_n \cos\phi_n \qquad b_n = k_n \sin\phi_n$$
$$\therefore\ k_n = \sqrt{a_n^2 + b_n^2} \qquad \phi_n = \arctan\frac{b_n}{a_n}$$

Now
$$\cos\theta + \mathrm{j}\sin\theta = \mathrm{e}^{\mathrm{j}\theta} \qquad \cos\theta - \mathrm{j}\sin\theta = \mathrm{e}^{-\mathrm{j}\theta}$$
$$\therefore\ \cos\theta = \frac{1}{2}(\mathrm{e}^{\mathrm{j}\theta} + \mathrm{e}^{-\mathrm{j}\theta})$$

$$f(x) = \tfrac{1}{2}a_0 + \sum_{n=1}^{\infty}\left\{k_n \frac{1}{2}\left[\mathrm{e}^{\mathrm{j}(nx-\phi_n)} + \mathrm{e}^{-\mathrm{j}(nx-\phi_n)}\right]\right\}$$
$$= \tfrac{1}{2}a_0 + \sum_{n=1}^{\infty}\left\{k_n \frac{1}{2}\left[\mathrm{e}^{\mathrm{j}nx}\mathrm{e}^{-\mathrm{j}\phi_n} + \mathrm{e}^{-\mathrm{j}nx}\mathrm{e}^{\mathrm{j}\phi_n}\right]\right\}$$

The term corresponding to $n = 0$ is the constant term of the series.

Substituting $n = 0$ in $k_n \frac{1}{2}[\mathrm{e}^{\mathrm{j}nx}\mathrm{e}^{-\mathrm{j}\phi_n} + \mathrm{e}^{-\mathrm{j}nx}\mathrm{e}^{\mathrm{j}\phi_n}]$

gives $k_0 \frac{1}{2}(\mathrm{e}^{-\mathrm{j}\phi_0} + \mathrm{e}^{\mathrm{j}\phi_0})$; $\qquad \phi_0 = -\phi_0 = 0.$

Therefore, this term $= k_0 \frac{1}{2}(2) = k_0\ (= \tfrac{1}{2}a_0)$.

$$f(x) = \sum_{n=0}^{\infty} k_n \frac{1}{2}[\mathrm{e}^{\mathrm{j}nx}\,\mathrm{e}^{-\mathrm{j}\phi_n} + \mathrm{e}^{-\mathrm{j}nx}\,\mathrm{e}^{\mathrm{j}\phi_n}] \qquad n = 0, 1, 2, \ldots$$

This can be written

$$f(x) = \sum_{n=-\infty}^{\infty} k_n\,\mathrm{e}^{\mathrm{j}nx}\,\mathrm{e}^{-\mathrm{j}\phi_n} \qquad n = 0, \pm 1, \pm 2, \ldots$$

where $k_{-n} = k_n$ and $\phi_{-n} = -\phi_n$.

$$\therefore\ f(x) = \sum_{n=-\infty}^{\infty} C_n\,\mathrm{e}^{\mathrm{j}nx} \tag{8.8}$$

where $C_n = k_n\,\mathrm{e}^{-\mathrm{j}\phi_n}$, i.e. a complex coefficient.

To determine C_n multiply both sides of (8.8) by $\mathrm{e}^{-\mathrm{j}mx}$ and integrate from $x = -\pi$ to $x = \pi$.

$$\int_{-\pi}^{\pi} f(x)\ \mathrm{e}^{-\mathrm{j}mx}\,\mathrm{d}x = \sum_{n=-\infty}^{\infty} C_n \int_{-\pi}^{\pi} \mathrm{e}^{\mathrm{j}nx}\ \mathrm{e}^{-\mathrm{j}mx}\,\mathrm{d}x$$

$$= \sum_{n=-\infty}^{\infty} C_n \int_{-\pi}^{\pi} e^{j(n-m)x}\, dx$$

$$= \sum_{n=-\infty}^{\infty} C_n \left[\frac{e^{j(n-m)x}}{j(n-m)}\right]_{-\pi}^{\pi} \qquad n \neq m$$

Let $(n-m) = A$, i.e. an integer. Then

$$\left[\frac{e^{j(n-m)x}}{j(n-m)}\right]_{-\pi}^{\pi} = \left[\frac{e^{jAx}}{jA}\right]_{-\pi}^{\pi} = \frac{1}{jA}\{e^{jA\pi} - e^{-jA\pi}\}$$

But $\quad e^{jA\pi} = \cos A\pi + j\sin A\pi$

and $\quad e^{-jA\pi} = \cos A\pi - j\sin A\pi$

$$\therefore\ e^{jA\pi} - e^{-jA\pi} = j2\sin A\pi = 0$$

For the case when $n = m$

$$\int_{-\pi}^{\pi} e^{j(n-m)x}\, dx = \int_{-\pi}^{\pi} 1\, dx = \Big[x\Big]_{-\pi}^{\pi} = 2\pi$$

$$\therefore \int_{-\pi}^{\pi} f(x)\, e^{-jnx}\, dx = C_n 2\pi \quad \therefore\ \underline{C_n = \frac{1}{2\pi}\int_{-\pi}^{\pi} f(x)\, e^{-jnx}\, dx}$$

Therefore, from (8.8) above

$$\underline{f(x) = \sum_{n=-\infty}^{\infty} C_n\, e^{jnx}}$$

where

$$\underline{C_n = \frac{1}{2\pi}\int_{-\pi}^{\pi} f(x)\, e^{-jnx}\, dx} \qquad n = 0, \pm 1, \pm 2, \ldots \tag{8.9}$$

Note that the complex coefficient C_n provides both the amplitude and the phase of the nth harmonic.

$$\begin{aligned} C_n &= k_n\, e^{-j\phi_n} = k_n(\cos\phi_n - j\sin\phi_n) \\ &= k_n\cos\phi_n - jk_n\sin\phi_n \\ &= a_n + jb_n \end{aligned}$$

where $\quad a_n = k_n\cos\phi_n$

$\qquad\quad b_n = -k_n\sin\phi_n$

Then

$$\text{amplitude} = \sqrt{a_n^2 + b_n^2}$$

$$\text{phase} = \phi_n = \arctan\frac{b_n}{a_n}$$

(b) *Period 2L*

For a function $f(x)$ of period $2L$, defined over the interval $-L$ to L, the result (8.9) becomes

$$f(x) = \sum_{n=-\infty}^{\infty} C_n \, e^{jn\pi x/L}$$

where $$C_n = \frac{1}{2L}\int_{-L}^{L} f(x)\, e^{-jn\pi x/L}\, dx \qquad n = 0, \pm 1, \pm 2, \ldots$$

(c) *Period T*

For a function $f(t)$ of period T, i.e. $\omega = \dfrac{2\pi}{T}$, the corresponding result is

$$f(t) = \sum_{n=-\infty}^{\infty} C_n \, e^{jn\omega t}$$

where $$C_n = \frac{1}{T}\int_{0}^{T} f(t)\, e^{-jn\omega t}\, dt \qquad n = 0, \pm 1, \pm 2, \ldots$$

Example 1

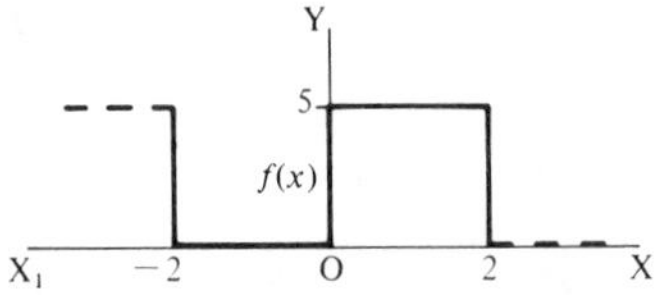

$$f(x) = 0 \qquad -2 < x < 0$$
$$f(x) = 5 \qquad 0 < x < 2$$
$$f(x) = f(x+4), \quad \text{i.e.} \quad L = 2.$$

$$f(x) = \sum_{n=-\infty}^{\infty} C_n \, e^{jn\pi x/L} \quad \text{where} \quad C_n = \frac{1}{2L}\int_{-L}^{L} f(x)\, e^{-jn\pi x/L}\, dx$$

$$C_n = \frac{1}{4}\int_{-2}^{2} f(x)\, e^{-jn\pi x/2}\, dx$$

$$\therefore\ 4C_n = \int_{0}^{2} 5e^{-jn\pi x/2}\, dx = \left[\frac{5e^{-jn\pi x/2}}{-jn\pi/2}\right]_0^2$$

$$= \frac{5(e^{-jn\pi} - 1)}{-jn\pi/2} = \frac{j10}{n\pi}(e^{-jn\pi} - 1)$$

But $$e^{-jn\pi} = \cos n\pi - j\sin n\pi = \left.\begin{array}{ll} 1 & \text{for } n \text{ even} \\ -1 & \text{for } n \text{ odd} \end{array}\right\} = (-1)^n$$

$$\therefore\ C_n = \frac{j5}{2n\pi}\{(-1)^n - 1\}$$

$$\therefore\ f(x) = \sum_{n=-\infty}^{\infty} \frac{j5}{2n\pi}\{(-1)^n - 1\}\, e^{jn\pi x/2} \qquad n \neq 0$$

For $n = 0$,

$$C_0 = \frac{1}{4}\int_{-2}^{2} f(x)\ e^0\, dx = \frac{1}{4}\int_0^2 5\, dx = \left[\frac{5x}{4}\right]_0^2 = \frac{5}{2}$$

Example 2

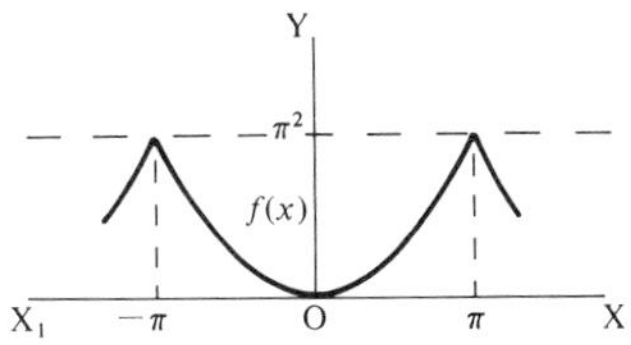

Y π^2 $f(x)$ X_1 $-\pi$ O π X

$$f(x) = x^2 \qquad -\pi < x < \pi$$
$$f(x) = f(x + 2\pi)$$

$$f(x) = \sum_{n=-\infty}^{\infty} C_n e^{jnx} \quad \text{where} \quad C_n = \frac{1}{2\pi}\int_{-\pi}^{\pi} f(x)\ e^{-jnx}\, dx$$

$$2\pi C_n = \int_{-\pi}^{\pi} x^2\ e^{-jnx}\, dx$$

$$= \left[x^2\left(\frac{e^{-jnx}}{-jn}\right)\right]_{-\pi}^{\pi} + \frac{2}{jn}\int_{-\pi}^{\pi} e^{-jnx}\, x\, dx$$

$$= \left[x^2\left(\frac{e^{-jnx}}{-jn}\right)\right]_{-\pi}^{\pi} + \frac{2}{jn}\left\{\left[x\left(\frac{e^{-jnx}}{-jn}\right)\right]_{-\pi}^{\pi} + \frac{1}{jn}\int_{-\pi}^{\pi} e^{-jnx}\, dx\right\}$$

$$= \left[x^2\,\frac{e^{-jnx}}{-jn}\right]_{-\pi}^{\pi} + \frac{2}{jn}\left\{\left[x\,\frac{e^{-jnx}}{-jn}\right]_{-\pi}^{\pi} + \frac{1}{jn}\left[\frac{e^{-jnx}}{-jn}\right]_{-\pi}^{\pi}\right\}$$

Now $\qquad e^{jn\pi} = e^{-jn\pi} = \cos n\pi = 1 \quad (n \text{ even})$

$$= -1 \quad (n \text{ odd})$$

The first and third terms are therefore zero.

$$2\pi C_n = \frac{2}{n^2}\{\pi e^{-jn\pi} + \pi e^{jn\pi}\} = \frac{4\pi}{n^2}(-1)^n$$

$$\therefore\ C_n = \frac{2}{n^2}(-1)^n$$

$$\therefore\ f(x) = \sum_{n=-\infty}^{\infty} \frac{2}{n^2}(-1)^n\, e^{jnx}$$

When $n = 0$,

$$C_0 = \frac{1}{2\pi}\int_{-\pi}^{\pi} x^2\,dx = \frac{1}{2\pi}\left(\frac{2\pi^3}{3}\right) \quad \therefore \; C_0 = \frac{\pi^2}{3}$$

8.4.3 Equivalence of the complex and trigonometrical forms

(a) *Trigonometrical form*

$$f(x) = \tfrac{1}{2}a_0 + \sum_{n=1}^{\infty}\{a_n \cos nx + b_n \sin nx\}$$

where

$$a_0 = \frac{1}{\pi}\int_{-\pi}^{\pi} f(x)\,dx$$

$$a_n = \frac{1}{\pi}\int_{-\pi}^{\pi} f(x)\cos nx\,dx$$

$$b_n = \frac{1}{\pi}\int_{-\pi}^{\pi} f(x)\sin nx\,dx$$

(b) *Complex form*

$$f(x) = \sum_{n=-\infty}^{\infty} C_n\,e^{jnx}$$

where

$$C_n = \frac{1}{2\pi}\int_{-\pi}^{\pi} f(x)\,e^{-jnx}\,dx$$

Now

$$C_n = \frac{1}{2\pi}\int_{-\pi}^{\pi} f(x)\,(\cos nx - j\sin nx)\,dx$$

$$= \frac{1}{2\pi}\int_{-\pi}^{\pi} f(x)\cos nx\,dx - j\frac{1}{2\pi}\int_{-\pi}^{\pi} f(x)\sin nx\,dx$$

$$\therefore\; C_n = \tfrac{1}{2}a_n - j\tfrac{1}{2}b_n \qquad (8.10)$$

Also

$$C_{-n} = \frac{1}{2\pi}\int_{-\pi}^{\pi} f(x)\{\cos(-nx) - j\sin(-nx)\}\,dx$$

$$= \frac{1}{2\pi}\int_{-\pi}^{\pi} f(x)\cos nx\,dx + j\frac{1}{2\pi}\int_{-\pi}^{\pi} f(x)\sin nx\,dx$$

$$\therefore\; C_{-n} = \tfrac{1}{2}a_n + j\tfrac{1}{2}b_n \qquad (8.11)$$

and $$C_0 = \frac{1}{2\pi}\int_{-\pi}^{\pi} f(x)\{\cos 0 - \mathrm{j}\sin 0\}\,\mathrm{d}x$$

$$= \frac{1}{2\pi}\int_{-\pi}^{\pi} f(x)\,\mathrm{d}x = \tfrac{1}{2}a_0 \quad \therefore \underline{C_0 = \tfrac{1}{2}a_0} \tag{8.12}$$

$$\therefore f(x) = \sum_{n=-\infty}^{\infty} C_n\,\mathrm{e}^{\mathrm{j}nx}$$

$$= C_0 + \sum_{n=1}^{\infty} C_n\,\mathrm{e}^{\mathrm{j}nx} + \sum_{n=1}^{\infty} C_{-n}\,\mathrm{e}^{-\mathrm{j}nx}$$

Substituting the expressions obtained above

$$f(x) = \tfrac{1}{2}a_0 + \sum_{n=1}^{\infty} \tfrac{1}{2}(a_n - \mathrm{j}b_n)\,\mathrm{e}^{\mathrm{j}nx} + \sum_{n=1}^{\infty} \tfrac{1}{2}(a_n + \mathrm{j}b_n)\,\mathrm{e}^{-\mathrm{j}nx}$$

$$= \tfrac{1}{2}a_0 + \sum_{n=1}^{\infty} \frac{a_n}{2}(\mathrm{e}^{\mathrm{j}nx} + \mathrm{e}^{-\mathrm{j}nx}) - \sum_{n=1}^{\infty} \mathrm{j}\frac{b_n}{2}(\mathrm{e}^{\mathrm{j}nx} - \mathrm{e}^{-\mathrm{j}nx})$$

$$= \tfrac{1}{2}a_0 + \sum_{n=1}^{\infty} a_n \cos nx + \sum_{n=1}^{\infty} b_n \sin nx$$

since $$\cos\theta = \frac{\mathrm{e}^{\mathrm{j}\theta} + \mathrm{e}^{-\mathrm{j}\theta}}{2}$$

and $$\sin\theta = \frac{\mathrm{e}^{\mathrm{j}\theta} - \mathrm{e}^{-\mathrm{j}\theta}}{\mathrm{j}2}$$

$$\therefore \underline{f(x) = \tfrac{1}{2}a_0 + \sum_{n=1}^{\infty}\{a_n \cos nx + b_n \sin nx\}}$$

Exercise 23

For each of the following functions, determine the complex form of the relevant Fourier series.

1. $f(x) = x \qquad -\pi < x < \pi$

   $f(x) = f(x + 2\pi)$

2. $f(x) = \mathrm{e}^x \qquad -1 < x < 1$

   $f(x) = f(x + 2)$

3. $f(t) = 1 - \mathrm{e}^{-t} \qquad 0 < t < 2$

   $f(t) = f(t + 2)$

4. $f(x) = \pi \qquad -\pi < x < 0$

$f(x) = x \qquad 0 < x < \pi$

$f(x) = f(x + 2\pi)$

5. $f(x) = x(x + 2\pi) \qquad 0 < x < 2\pi$

$f(x) = f(x + 2\pi)$

6. $f(t) = A \sin \omega t \qquad 0 < t < \frac{\pi}{\omega}$

$f(t) = 0 \qquad \frac{\pi}{\omega} < t < \frac{2\pi}{\omega}$

$f(t) = f\left(t + \frac{2\pi}{\omega}\right)$

7. $f(x) = 0 \qquad -\pi < x < 0$

$f(x) = e^x - 1 \qquad 0 < x < \pi$

$f(x) = f(x + 2\pi)$

8.5 REVISION SUMMARY

1. Integration and differentiation of a Fourier series

A Fourier series can always be integrated term by term.

A Fourier series can be differentiated term by term only if certain conditions are satisfied.

2. Gibbs' phenomenon

Overshoot immediately before and after a point of discontinuity.

3. Fourier coefficients from jumps at discontinuities

$$f(x) = \tfrac{1}{2}a_0 + \sum_{n=1}^{\infty} \{a_n \cos nx + b_n \sin nx\} \qquad \text{Period } 2\pi$$

$$a_0 = \frac{1}{\pi}\int_{-\pi}^{\pi} f(x)\,dx$$

$$a_n = \frac{1}{n\pi}\left\{-\sum_{r=1}^{r} \mathrm{j}_r \sin nx_r - \frac{1}{n}\sum_{r=1}^{r} \mathrm{j}_r' \cos nx_r + \frac{1}{n^2}\sum_{r=1}^{r} \mathrm{j}_r'' \sin nx_r + \frac{1}{n^3}\sum_{r=1}^{r} \mathrm{j}_r''' \cos nx_r - \ldots\right\}$$

$$b_n = \frac{1}{n\pi}\left\{\sum_{r=1}^{r} \mathrm{j}_r \cos nx_r - \frac{1}{n}\sum_{r=1}^{r} \mathrm{j}_r' \sin nx_r - \frac{1}{n^2}\sum_{r=1}^{r} \mathrm{j}_r'' \cos nx_r + \frac{1}{n^3}\sum_{r=1}^{r} \mathrm{j}_r''' \sin nx_r + \ldots\right\}$$

where, in each case, $n = 1, 2, 3, \ldots$, and where $j_3'' \equiv$ jump in the second derivative at the third discontinuity in the cycle.

4. Complex forms of Fourier series

(a) (i) Period 2π $\quad a_n + \mathrm{j}b_n = \dfrac{1}{\pi}\displaystyle\int_0^{2\pi} f(x)\,\mathrm{e}^{\mathrm{j}nx}\,\mathrm{d}x \qquad n = 1, 2, 3, \ldots$

(ii) Period $2L$ $\quad a_n + \mathrm{j}b_n = \dfrac{1}{L}\displaystyle\int_{-L}^{L} f(x)\,\mathrm{e}^{\mathrm{j}n\pi x/L}\,\mathrm{d}x$

(iii) Period T $\quad a_n + \mathrm{j}b_n = \dfrac{2}{T}\displaystyle\int_0^{T} f(t)\,\mathrm{e}^{\mathrm{j}n\omega t}\,\mathrm{d}t \qquad$ where $\omega = \dfrac{2\pi}{T}$.

(b) (i) Period 2π $\quad f(x) = \displaystyle\sum_{n=-\infty}^{\infty} C_n\,\mathrm{e}^{\mathrm{j}nx}; \qquad C_n = \dfrac{1}{2\pi}\displaystyle\int_{-\pi}^{\pi} f(x)\,\mathrm{e}^{-\mathrm{j}nx}\,\mathrm{d}x$

(ii) Period $2L$ $\quad f(x) = \displaystyle\sum_{n=-\infty}^{\infty} C_n\,\mathrm{e}^{\mathrm{j}n\pi x/L}; \qquad C_n = \dfrac{1}{2L}\displaystyle\int_{-L}^{L} f(x)\,\mathrm{e}^{-\mathrm{j}n\pi x/L}\,\mathrm{d}x$

(iii) Period T $\quad f(t) = \displaystyle\sum_{n=-\infty}^{\infty} C_n\,\mathrm{e}^{\mathrm{j}n\omega t}; \qquad C_n = \dfrac{1}{T}\displaystyle\int_0^{T} f(t)\,\mathrm{e}^{-\mathrm{j}n\omega t}\,\mathrm{d}t$

where $\omega = \dfrac{2\pi}{T}$.

In each case, $n = 0, \pm 1, \pm 2, \ldots$

Chapter 9

SOLUTION OF BOUNDARY VALUE PROBLEMS

9.1 ORDINARY SECOND ORDER LINEAR DIFFERENTIAL EQUATIONS (SUMMARY)

9.1.1 Equations of the form $a\dfrac{d^2y}{dx^2} + b\dfrac{dy}{dx} + cy = 0$

Auxiliary equation: $am^2 + bm + c = 0$

(a) Real and different roots, $m = m_1$ and $m = m_2$.

$$\text{Solution:} \quad y = Ae^{m_1x} + Be^{m_2x} \tag{9.1}$$

(b) Real and equal roots, $m = m_1$ (twice).

$$\text{Solution:} \quad y = e^{m_1x}(A + Bx) \tag{9.2}$$

(c) Complex roots, $m = \alpha \pm j\beta$.

$$\text{Solution:} \quad y = e^{\alpha x}(A\cos\beta x + B\sin\beta x) \tag{9.3}$$

9.1.2 Equations of the form $\dfrac{d^2y}{dx^2} \pm n^2y = 0$

(a) $\dfrac{d^2y}{dx^2} + n^2y = 0$

$$\text{Solution:} \quad y = A\cos nx + B\sin nx \tag{9.4}$$

(b) $\dfrac{d^2y}{dx^2} - n^2y = 0$

$$\text{Solution:} \quad y = A\cosh nx + B\sinh nx \tag{9.5}$$

$$\text{or} \quad y = Ae^{nx} + Be^{-nx}$$

$$\text{or} \quad y = A\sinh n(x + \phi)$$

In each of the above solutions, A and B are arbitrary constants.

9.1.3 Equations of the form $a\dfrac{d^2y}{dx^2} + b\dfrac{dy}{dx} + cy = f(x)$

(a) Determine the *complementary function*, i.e. solve the equation $a\dfrac{d^2y}{dx^2} + b\dfrac{dy}{dx} + cy = 0$ as above.

(b) Form a *particular integral*, i.e. assume the general form of $f(x)$; differentiate twice with respect to x; substitute in the left hand side; and equate coefficients. If the general form of $f(x)$ is already included in the complementary function, multiply by x and proceed as before.

(c) The general solution = complementary function + particular integral.

Note: In general, if $y = u_1, y = u_2, y = u_3, \ldots,$ are solutions of a linear equation, so also is $y = u_1 + u_2 + u_3 + \ldots$

9.2 PARTIAL DIFFERENTIAL EQUATIONS

A *partial differential equation* is a relationship between a dependent variable (u), two or more independent variables $(x, y, t, \ldots)$ and partial differential coefficients of u with respect to these independent variables. The solution of the equation is then expressed in the form $u = f(x, y, t, \ldots)$.

The arbitrary functions that arise from any differential equation are determined, in any particular case, by applying to the general solution the additional information supplied by the *initial conditions* or, in general, the *boundary conditions* since the information does not always relate to zero values of the independent variables.

Furthermore, if $u = u_1, u = u_2, u = u_3, \ldots$ are different solutions of a linear partial differential equation, so also is

$$u = c_1 u_1 + c_2 u_2 + c_3 u_3 + \ldots$$

where $c_1, c_2, c_3, \ldots,$ are arbitrary constants.

The study of partial differential equations can be extensive and we will restrict our attention to a number of such equations of types that occur in science and technology, that can be solved by the method of separating the variables, and that require some application of Fourier series techniques.

9.3 THE WAVE EQUATION

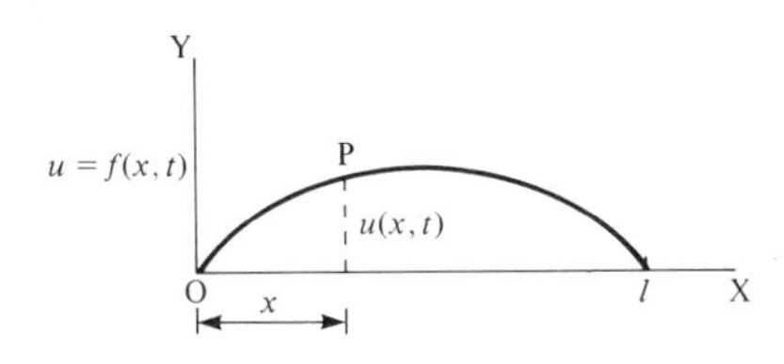

Consider a perfectly flexible elastic string, stretched between two points at $x = 0$ and $x = l$ with uniform tension T.

If the string is displaced slightly from its initial position while the ends remain fixed, and then released, the string will oscillate. The position of any point P in the string at any instant will then be a function of its distance from one end (x) of the string and also of time (t), i.e. $u = f(x, t)$. For convenience, the function can be denoted by $u(x, t)$.

The equation of motion is given by the partial differential equation

$$\frac{\partial^2 u}{\partial x^2} = \frac{1}{c^2}\frac{\partial^2 u}{\partial t^2}$$

where $$c^2 = \frac{T}{\rho} = \frac{\text{tension}}{\text{mass per unit length of the string}}$$

9.3.1 Solution of the wave equation

Equation of motion:

$$\frac{\partial^2 u}{\partial x^2} = \frac{1}{c^2}\frac{\partial^2 u}{\partial t^2} \qquad u = f(x,t)$$

Boundary conditions:

(a) String fixed at both ends, i.e. at $x = 0$ and at $x = l$

$$\left.\begin{aligned} \therefore\ u(0,t) &= 0 \\ u(l,t) &= 0 \end{aligned}\right\} \quad \text{for all values of } t \geqslant 0$$

Initial conditions:

(b) Let the initial deflection (i.e. at $t = 0$) at any point P distant x from one end be $f(x)$

$$\therefore\ u(x,0) = f(x)$$

(c) Let the initial velocity at P be $g(x)$

i.e. $$\left[\frac{\partial u}{\partial t}\right]_{t=0} = g(x)$$

Separating the variables

Assume that the solution is of the form $u(x,t) = X(x)T(t)$

where X is a function of x only

T is a function of t only

$$u = XT \quad \therefore \frac{\partial u}{\partial x} = X'T \quad \text{and} \quad \frac{\partial^2 u}{\partial x^2} = X''T$$

$$\frac{\partial u}{\partial t} = XT' \quad \text{and} \quad \frac{\partial^2 u}{\partial t^2} = XT''$$

where, in each case, the primes denote differential coefficients with respect to its own independent variable, e.g. $X'' = \dfrac{d^2X}{dx^2}$.

Substituting in $\dfrac{\partial^2 u}{\partial x^2} = \dfrac{1}{c^2}\dfrac{\partial^2 u}{\partial t^2}$, we have

$$X''T = \frac{1}{c^2}XT'' \qquad \therefore\ XT'' = c^2X''T$$

$$\therefore\ \frac{X''}{X} = \frac{1}{c^2}\frac{T''}{T}$$

The left hand expression involves functions of x only.

The right hand expression involves functions of t only.

Therefore, to be equal, both expressions must be equal to a constant (k).

$$\therefore\ \frac{X''}{X} = k \quad \text{and} \quad \frac{1}{c_2}\frac{T''}{T} = k$$

$$\therefore\ X'' - kX = 0 \qquad T'' - c^2kT = 0$$

k is an arbitrary constant. What do we know about it?

(a) *If* $k = 0$, $X'' = 0 \ \therefore\ X' = a \ \therefore\ X = ax + b$

But

$X = 0$ at $x = 0 \ \therefore\ b = 0 \ \therefore\ X = ax$

and $X = 0$ at $x = l \ \therefore\ a = 0$

$\therefore\ a = b = 0 \ \therefore\ \underline{X = 0}$

(b) *If k is positive*, i.e. $k = p^2$, $X'' - p^2X = 0$.

$$\therefore\ m^2 - p^2m = 0 \qquad \therefore\ m = 0 \ \text{or} \ m = p^2 \quad \therefore\ X = A + Be^{p^2x}$$

But $X = 0$ at $x = 0 \quad \therefore\ 0 = A + B \qquad \therefore\ B = -A$

and $X = 0$ at $x = l \quad \therefore\ 0 = A + Be^{p^2l} \qquad \therefore\ 0 = A(1 - e^{p^2l})$

$\therefore\ A = 0$ or $e^{p^2l} = 1$, i.e. $p^2l = 0 \ \therefore\ A = B = 0 \ \therefore\ \underline{X = 0}$

(c) *If k is negative*, i.e. $k = -p^2$, $X'' + p^2X = 0$.

This gives the solution $X = A\cos px + B\sin px$.

Similarly, for $T'' - c^2kT = 0$, putting $k = -p^2$

$$T'' + c^2p^2T = 0$$

This gives the solution $T = C\cos pt + D\sin cpt$.

$$\therefore\ u(x,t) = (A\cos px + B\sin px)(C\cos cpt + D\sin cpt)$$

Let $cp = \lambda \ \therefore\ p = \dfrac{\lambda}{c}$

$$\therefore\ \underline{u(x,t) = \left(A\cos\frac{\lambda}{c}x + B\sin\frac{\lambda}{c}x\right)(C\cos\lambda t + D\sin\lambda t)} \tag{9.6}$$

where A, B, C, D are arbitrary constants. This is the general solution.

The result must also satisfy the boundary conditions.

(a_1) $u = 0$ when $x = 0$ for all t

$$\therefore\ 0 = A(C\cos\lambda t + D\sin\lambda t) \qquad \therefore\ \underline{A = 0}$$

$$\therefore\ u(x,t) = B\sin\frac{\lambda}{c}x(C\cos\lambda t + D\sin\lambda t)$$

(a_2) $u = 0$ when $x = l$ for all t

$$\therefore\ 0 = B\sin\frac{\lambda}{c}l(C\cos\lambda t + D\sin\lambda t)$$

Now B cannot be zero or the function $u(x,t)$ would be identically zero.

$$\therefore\ \sin\frac{\lambda}{c}l = 0 \qquad \therefore\ \frac{\lambda}{c}l = n\pi \qquad n = 1, 2, 3, \ldots$$

$$\therefore\ \lambda = \frac{nc\pi}{l} \qquad n = 1, 2, 3, \ldots$$

The value $n = 0$ is excluded since this would make $u(x,t)$ identically zero.

Note: There is thus an infinite set of values of λ and each discrete value of λ gives a particular solution for $u(x,t)$. The values of λ are called the *eigenvalues* and each particular solution the corresponding *eigenfunction*.

We therefore have:

n	$\lambda = \dfrac{nc\pi}{l}$	$u(x,t) = B\sin\dfrac{\lambda x}{c}\{C\cos\lambda t + D\sin\lambda t\}$
1	$\lambda_1 = \dfrac{c\pi}{l}$	$u_1 = \sin\dfrac{\pi x}{l}\left\{C_1\cos\dfrac{c\pi t}{l} + D_1\sin\dfrac{c\pi t}{l}\right\}$
2	$\lambda_2 = \dfrac{2c\pi}{l}$	$u_2 = \sin\dfrac{2\pi x}{l}\left\{C_2\cos\dfrac{2c\pi t}{l} + D_2\sin\dfrac{2c\pi t}{l}\right\}$
3	$\lambda_3 = \dfrac{3c\pi}{l}$	$u_3 = \sin\dfrac{3\pi x}{l}\left\{C_3\cos\dfrac{3c\pi t}{l} + D_3\sin\dfrac{3c\pi t}{l}\right\}$
$\vdots$	$\vdots$	$\vdots$
r	$\lambda_r = \dfrac{rc\pi}{l}$	$u_r = \sin\dfrac{r\pi x}{l}\left\{C_r\cos\dfrac{rc\pi t}{l} + D_r\sin\dfrac{rc\pi t}{l}\right\}$
$\vdots$	$\vdots$	$\vdots$

where $C_1, C_2, C_3, \ldots,$ and $D_1, D_2, D_3, \ldots,$ are arbitrary constants.

Since the given equation

$$\frac{\partial^2 u}{\partial x^2} = \frac{1}{c^2}\frac{\partial^2 u}{\partial t^2}$$

is linear in form, then if $u = u_1, u = u_2, u = u_3, \ldots$ are particular solutions, a more general solution is $u = u_1 + u_2 + u_3 + \ldots$

The general solution is therefore

$$u(x,t) = \sum_{r=1}^{\infty} u_r = \sum_{r=1}^{\infty}\left\{\sin\frac{r\pi x}{l}\left(C_r\cos\frac{rc\pi t}{l} + D_r\sin\frac{rc\pi t}{l}\right)\right\} \tag{9.7}$$

To find C_r and D_r, we apply the remaining boundary conditions, i.e. the initial conditions.

(b) At $t = 0$, $u(x,0) = f(x)$, $0 \leqslant x \leqslant l$.

$$\therefore\ u(x,0) = f(x) = \sum_{r=1}^{\infty} C_r\sin\frac{r\pi x}{l}$$

(c) Also at $t = 0$, $\left[\frac{\partial u(x,t)}{\partial t}\right]_{t=0} = g(x)$, $0 \leqslant x \leqslant l$.

Differentiate (9.7) with respect to t and put $t = 0$

$$\frac{\partial u}{\partial t} = \sum_{r=1}^{\infty}\sin\frac{r\pi x}{l}\left\{-C_r\frac{rc\pi}{l}\sin\frac{rc\pi t}{l} + D_r\frac{rc\pi}{l}\cos\frac{rc\pi t}{l}\right\}$$

When $t = 0$,

$$\frac{\partial u}{\partial t} = g(x) = \sum_{r=1}^{\infty} D_r\frac{rc\pi}{l}\sin\frac{r\pi x}{l}$$

$$\therefore\ g(x) = \frac{c\pi}{l}\sum_{r=1}^{\infty} r\sin\frac{r\pi x}{l}$$

The coefficients C_r and D_r can now be found by applying Fourier series techniques.

$$C_r = 2\times\text{mean value of } f(x)\sin\frac{r\pi x}{l} \quad \text{from} \quad x = 0 \quad \text{to} \quad x = l.$$

$$\therefore\ C_r = \frac{2}{l}\int_0^l f(x)\sin\frac{r\pi x}{l}\,dx \qquad r = 1, 2, 3, \ldots$$

Similarly,

$$\frac{rc\pi}{l}D_r = 2\times\text{mean value of } g(x)\sin\frac{r\pi x}{l} \quad \text{from} \quad x = 0 \quad \text{to} \quad x = l.$$

$$\therefore\ D_r = \frac{2}{rc\pi}\int_0^l g(x)\sin\frac{r\pi x}{l}\,dx \qquad r = 1,2,3,\ldots$$

Substituting the expressions for C_r and D_r in the general solution (9.7) gives

$$u(x,t) = \sum_{r=1}^{\infty}\left\{\left[\frac{2}{l}\int_0^l f(w)\sin\frac{r\pi w}{l}\,dw\right]\cos\frac{rc\pi t}{l}\sin\frac{r\pi x}{l} + \left[\frac{2}{rc\pi}\int_0^l g(w)\sin\frac{r\pi w}{l}\,dw\right]\sin\frac{rc\pi t}{l}\sin\frac{r\pi x}{l}\right\}$$

where the variable w has been introduced in the definite integrals to distinguish it from the independent variable x of the function $u(x,t)$.

This is now the particular solution satisfying the boundary conditions stated.

Example

The centre of an elastic string 4 m long is deflected initially to P as shown and then released. The initial velocity is zero. Determine the solution of the wave equation $\dfrac{\partial^2 u}{\partial x^2} = \dfrac{1}{c^2}\dfrac{\partial^2 u}{\partial t^2}$ where $c^2 = 100$.

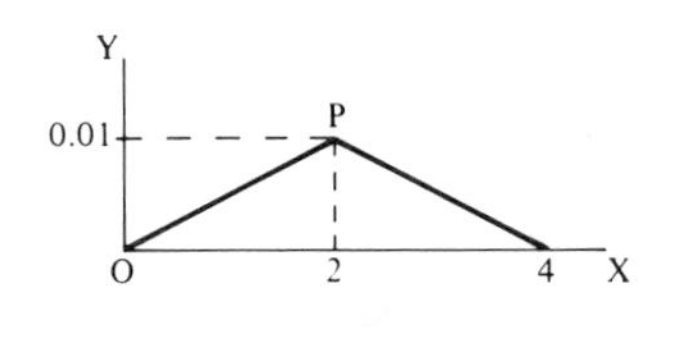

At $t = 0$,

$$f(x) = \frac{x}{200} \qquad 0 < x < 2$$

$$f(x) = \frac{2}{100} - \frac{x}{200} \qquad 2 < x < 4$$

(a) Boundary conditions:

$$\left.\begin{aligned} u(0,t) &= 0 \\ u(4,t) &= 0 \end{aligned}\right\} \text{ for all } t \geqslant 0$$

(b) Initial conditions:

$$u(x,0) = f(x) = \frac{x}{200} \qquad 0 < x < 2$$

$$= \frac{4-x}{200} \qquad 2 < x < 4$$

$$\left[\frac{\partial u(x,t)}{\partial t}\right]_{t=0} = g(x) = 0 \qquad 0 < x < 4$$

As before, we use the method of separating the variables. Let

$$u(x,t) = X(x)T(t) \quad \text{where} \quad X \text{ is a function of } x \text{ only}$$
$$\text{and} \quad T \text{ is a function of } t \text{ only}$$

$$u = XT \quad \therefore \frac{\partial u}{\partial x} = X'T \qquad \frac{\partial^2 u}{\partial x^2} = X''T$$

$$\frac{\partial u}{\partial t} = XT' \qquad \frac{\partial^2 u}{\partial t^2} = XT''$$

$$\frac{\partial^2 u}{\partial x^2} = \frac{1}{c^2}\frac{\partial^2 u}{\partial t^2} \qquad \therefore X''T = \frac{1}{c^2}XT''$$

$$\therefore \frac{X''}{X} = \frac{1}{c^2}\frac{T''}{T}$$

Left hand side is a function of x only; Right hand side is a function of t only — To be equal, they must both equal a constant (k).

As in the previous working, for an oscillatory solution, put $k = -p^2$.

$$\therefore \frac{X''}{X} = -p^2 \qquad \therefore X'' + p^2X = 0 \quad \therefore X = A\cos px + B\sin px$$

$$\text{and} \quad \frac{1}{c^2}\frac{T''}{T} = -p^2 \qquad \therefore T'' + c^2p^2T = 0 \quad \therefore T = C\cos cpt + D\sin cpt$$
$$= C\cos 10pt + D\sin 10pt$$

$$\therefore \underline{u(x,t) = XT = \{A\cos px + B\sin px\}\{C\cos 10pt + D\sin 10pt\}} \qquad (9.8)$$

where A, B, C, D are arbitrary constants.

From the boundary conditions,

(a) $$u = 0, \quad x = 0 \quad \text{for all} \quad t \geqslant 0$$

$$\therefore 0 = A\{C\cos 10pt + D\sin 10pt\} \qquad \therefore \underline{A = 0}$$

$$\therefore u(x,t) = B\sin px\{C\cos 10pt + D\sin 10pt\} \qquad (9.9)$$

(b) $$u = 0, \quad x = 4 \quad \text{for all} \quad t \geqslant 0$$

$$\therefore 0 = B\sin 4p\{C\cos 10pt + D\sin 10pt\}$$

$$B \neq 0 \quad \therefore \sin 4p = 0 \qquad \therefore 4p = n\pi \qquad n = 1, 2, 3, \ldots$$

$$\therefore p = \frac{n\pi}{4} \qquad n = 1, 2, 3, \ldots$$

n	$p = \frac{n\pi}{4}$	$u(x,t) = B\sin\frac{n\pi x}{4}\left\{C\cos\frac{5n\pi t}{2} + D\sin\frac{5n\pi t}{2}\right\}$
1	$p_1 = \frac{\pi}{4}$	$u_1 = \sin\frac{\pi x}{4}\left\{C_1\cos\frac{5\pi t}{2} + D_1\sin\frac{5\pi t}{2}\right\}$
2	$p_2 = \frac{\pi}{2}$	$u_2 = \sin\frac{\pi x}{2}\left\{C_2\cos\frac{10\pi t}{2} + D_2\sin\frac{10\pi t}{2}\right\}$
3	$p_3 = \frac{3\pi}{4}$	$u_3 = \sin\frac{3\pi x}{4}\left\{C_3\cos\frac{15\pi t}{2} + D_3\sin\frac{15\pi t}{2}\right\}$
.	.	.
.	.	.
.	.	.
r	$p_r = \frac{r\pi}{4}$	$u_r = \sin\frac{r\pi x}{4}\left\{C_r\cos\frac{5r\pi t}{2} + D_r\sin\frac{5r\pi t}{2}\right\}$
.	.	.
.	.	.
.	.	.

where, for each expression, $C_r = (B\times C)_r$ and $D_r = (B\times D)_r$.

$$u = u_1+u_2+u_3+\ldots+u_r+\ldots$$

$$\therefore\ u(x,t) = \sum_{r=1}^{\infty}\left\{\sin\frac{r\pi x}{4}\left(C_r\cos\frac{5r\pi t}{2} + D_r\sin\frac{5r\pi t}{2}\right)\right\} \qquad (9.10)$$

To find C_r and D_r we use the initial conditions.

(a) At $t = 0$,

$$u(x,0) = f(x) = \frac{x}{200} \qquad 0<x<2$$

$$= \frac{4-x}{200} \qquad 2<x<4$$

But, from (9.10),

$$u(x,0) = \sum_{r=1}^{\infty} C_r\sin\frac{r\pi x}{4} \qquad (9.11)$$

(b) Also at $t = 0$, $\frac{\partial u}{\partial t} = 0$. Differentiating (9.10) with respect to t,

$$\frac{\partial u}{\partial t} = \sum_{r=1}^{\infty}\left\{\sin\frac{r\pi x}{4}\left(-C_r\frac{5r\pi}{2}\sin\frac{5r\pi t}{2} + D_r\frac{5r\pi}{2}\cos\frac{5r\pi t}{2}\right)\right\}$$

$$\left[\frac{\partial u}{\partial t}\right]_{t=0} = g(x) = 0 \qquad \therefore\ 0 = \sum_{r=1}^{\infty}\left\{\sin\frac{r\pi x}{4}D_r\frac{5r\pi}{2}\right\} \qquad (9.12)$$

Applying Fourier methods to (9.11),

$$C_r = 2\times\text{mean value of } f(x)\sin\frac{r\pi x}{4} \quad \text{from} \quad x = 0 \text{ to } x = 4$$

$$= \frac{2}{4}\int_0^4 f(x)\sin\frac{r\pi x}{4}\,\mathrm{d}x \qquad r = 1,2,3,\ldots$$

$$\therefore\ 2C_r = \int_0^2 \frac{x}{200}\sin\frac{r\pi x}{4}\,\mathrm{d}x + \int_2^4 \frac{4-x}{200}\sin\frac{r\pi x}{4}\,\mathrm{d}x$$

$$= \qquad I_1 \qquad + \qquad I_2$$

$$I_1 = \frac{1}{200}\left\{\left[x\left(-\frac{4}{r\pi}\cos\frac{r\pi x}{4}\right)\right]_0^2 + \frac{4}{r\pi}\int_0^2 \cos\frac{r\pi x}{4}\,\mathrm{d}x\right\}$$

$$= \frac{1}{200}\left\{-\frac{8}{r\pi}\cos\frac{r\pi}{2} + \frac{4^2}{r^2\pi^2}\left[\sin\frac{r\pi x}{4}\right]_0^2\right\}$$

$$= \frac{1}{200}\left\{-\frac{8}{r\pi}\cos\frac{r\pi}{2} + \frac{4^2}{r^2\pi^2}\sin\frac{r\pi}{2}\right\}$$

$$I_2 = \frac{1}{200}\left\{\left[(4-x)\left(-\frac{4}{r\pi}\cos\frac{r\pi x}{4}\right)\right]_2^4 - \frac{4}{r\pi}\int_2^4 \cos\frac{r\pi x}{4}\,\mathrm{d}x\right\}$$

$$= \frac{1}{200}\left\{\frac{8}{r\pi}\cos\frac{r\pi}{2} - \frac{4}{r\pi}\left[\frac{4}{r\pi}\sin\frac{r\pi x}{4}\right]_2^4\right\}$$

$$= \frac{1}{200}\left\{\frac{8}{r\pi}\cos\frac{r\pi}{2} - \frac{4^2}{r^2\pi^2}\left(\sin r\pi - \sin\frac{r\pi}{2}\right)\right\}$$

$$\therefore\ 2C_r = \frac{1}{200}\left\{-\frac{8}{r\pi}\cos\frac{r\pi}{2} + \frac{4^2}{r^2\pi^2}\sin\frac{r\pi}{2} + \frac{8}{r\pi}\cos\frac{r\pi}{2} - \frac{4^2}{r^2\pi^2}\sin r\pi + \frac{4^2}{r^2\pi^2}\sin\frac{r\pi}{2}\right\} \qquad r = 1,2,3,\ldots$$

$$= \frac{1}{200}\left\{\frac{2\times 4^2}{r^2\pi^2}\sin\frac{r\pi}{2}\right\} \qquad \text{since} \quad \sin r\pi = 0$$

$$\therefore\ C_r = \frac{4^2}{200\pi^2}\left\{\frac{1}{r^2}\sin\frac{r\pi}{2}\right\} \qquad r = 1,2,3,\ldots$$

Similarly, from (9.12)

$$\frac{5r\pi}{2}D_r = 2\times\text{the mean value of } g(x)\sin\frac{r\pi x}{4} \quad \text{from} \quad x = 0 \text{ to } x = 4$$

$$= \frac{2}{4}\int_0^4 g(x)\sin\frac{r\pi x}{4}\,\mathrm{d}x$$

But

$$g(x) = 0 \qquad \therefore \underline{D_r = 0}$$

$$\therefore u(x,t) = \sum_{r=1}^{\infty}\left\{\sin\frac{r\pi x}{4} C_r \cos\frac{5r\pi t}{2}\right\}$$

$$\underline{u(x,t) = \frac{4^2}{200\pi^2}\sum_{r=1}^{\infty}\left\{\frac{1}{r^2}\sin\frac{r\pi}{2}\sin\frac{r\pi x}{4}\cos\frac{5r\pi t}{2}\right\}}$$

$$\therefore u(x,t) = \frac{4^2}{200\pi^2}\left\{\frac{1}{1^2}\sin\frac{\pi x}{4}\cos\frac{5\pi t}{2} - \frac{1}{3^2}\sin\frac{3\pi x}{4}\cos\frac{15\pi t}{2} + \frac{1}{5^2}\sin\frac{5\pi x}{4}\cos\frac{25\pi t}{2} - \ldots\right\}$$

9.4 THE HEAT CONDUCTION EQUATION for a uniform finite bar

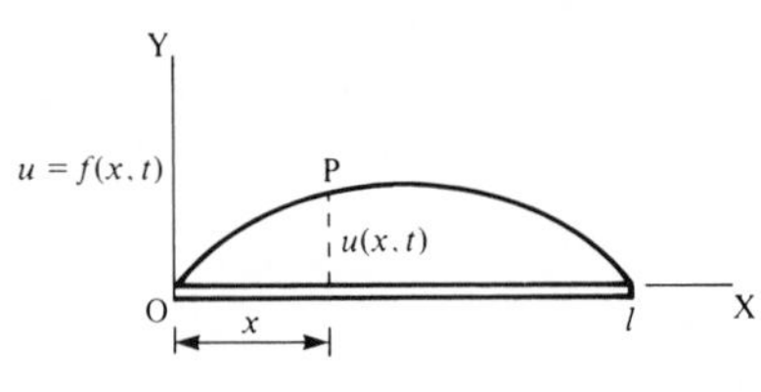

The flow of heat in a uniform bar depends on the distribution of temperature along the bar, on the thermal conductivity and specific heat of the material, and on the mass per unit length of the bar. Consider a bar, insulated throughout its length, except at its two ends, where the temperature u at any point is a function of its distance x from one end and the time t.

The relevant one-dimensional heat equation is then of the form

$$\frac{\partial^2 u}{\partial x^2} = \frac{1}{c^2}\frac{\partial u}{\partial t} \tag{9.13}$$

where $c^2 = \dfrac{k}{\sigma\rho}$ and k = thermal conductivity of the material

σ = specific heat of the material

ρ = mass per unit length of the bar

9.4.1 Solution of the heat conduction equation

Consider the case where

(a) the ends of the bar are at $x = 0$ and at $x = l$

(b) the ends of the bar are maintained at zero temperature

(c) the initial temperature distribution along the bar is defined by $f(x)$.

Boundary conditions:

$$\left.\begin{array}{ll}\text{(a)} & u(0,t)=0\\ \text{(b)} & u(l,t)=0\end{array}\right\}\qquad \text{for all } t\geqslant 0$$

Initial condition:

(c) $u(x,0)=f(x)$ $\qquad 0\leqslant x\leqslant l$

As before, we use the method of separating the variables and assume a solution in the form

$$u(x,t) = X(x)T(t)$$

where X is a function of x only

T is a function of t only.

$$u = XT \qquad \therefore \frac{\partial u}{\partial x} = X'T \qquad \frac{\partial^2 u}{\partial x^2} = X''T$$

$$\frac{\partial u}{\partial t} = XT'$$

$$\frac{\partial^2 u}{\partial x^2} = \frac{1}{c^2}\frac{\partial u}{\partial t} \qquad \therefore X''T = \frac{1}{c^2}XT'$$

$$\therefore \frac{X''}{X} = \frac{1}{c^2}\frac{T'}{T}$$

To be equal, these must each equal a constant $(-p^2)$

$$\therefore X''+p^2X = 0 \quad \text{giving} \quad X = A\cos px + B\sin px$$

$$T'+p^2c^2T = 0 \qquad \therefore \frac{T'}{T} = -p^2c^2 \qquad \therefore \ln T = -p^2c^2t+c$$

$$\therefore T = Ce^{-p^2c^2t}$$

$$\therefore u = XT = (A\cos px + B\sin px)Ce^{-p^2c^2t}$$

$$\underline{u(x,t) = (P\cos px + Q\sin px)e^{-p^2c^2t}} \qquad (9.14)$$

From the boundary conditions,

$$u(0,t) = 0 \qquad \therefore 0 = Pe^{-p^2c^2t} \quad \text{for all } t\geqslant 0 \qquad \therefore \underline{P = 0}$$

$$\therefore \underline{u(x,t) = Qe^{-p^2c^2t}\sin px} \qquad (9.15)$$

Also $\qquad u(l,t) = 0 \qquad \therefore 0 = Qe^{-p^2c^2t}\sin pl \quad$ for all $t\geqslant 0$

$Q\neq 0$ or $u(x,t)$ would be identically zero.

$$\therefore \sin pl = 0 \qquad \therefore pl = n\pi \qquad \therefore p = \frac{n\pi}{l}$$

$$n = 1,2,3,\ldots$$

n	$p = \dfrac{n\pi}{l}$	$u = Q\mathrm{e}^{-p^2c^2t}\sin px = Q\mathrm{e}^{-n^2c^2\pi^2t/l^2}\sin\dfrac{n\pi x}{l}$
1	$p_1 = \dfrac{\pi}{l}$	$u_1 = Q_1\mathrm{e}^{-c^2\pi^2t/l^2}\sin\dfrac{\pi x}{l}$
2	$p_2 = \dfrac{2\pi}{l}$	$u_2 = Q_2\mathrm{e}^{-4c^2\pi^2t/l^2}\sin\dfrac{2\pi x}{l}$
3	$p_3 = \dfrac{3\pi}{l}$	$u_3 = Q_3\mathrm{e}^{-9c^2\pi^2t/l^2}\sin\dfrac{3\pi x}{l}$
.	.	.
.	.	.
.	.	.
r	$p_r = \dfrac{r\pi}{l}$	$u_r = Q_r\mathrm{e}^{-n^2c^2\pi^2t/l^2}\sin\dfrac{r\pi x}{l}$
.	.	.
.	.	.
.	.	.

As before, $u = u_1 + u_2 + u_3 + \ldots + u_r + \ldots$

$$\therefore\ u(x,t) = \sum_{r=1}^{\infty}\left\{Q_r\mathrm{e}^{-r^2c^2\pi^2t/l^2}\sin\frac{r\pi x}{l}\right\} \qquad (9.16$$

When $t = 0$, $\quad u(x,0) = f(x) \qquad 0 \leqslant x \leqslant l$

$$\therefore\ u(x,0) = f(x) = \sum_{r=1}^{\infty}\left\{Q_r\sin\frac{r\pi x}{l}\right\}$$

and from Fourier series techniques,

$$Q_r = 2\times\text{mean value of } f(x)\sin\frac{r\pi x}{l} \quad \text{from} \quad x = Q \text{ to } x = l$$

$$= \frac{2}{l}\int_0^l f(x)\sin\frac{r\pi x}{l}\,\mathrm{d}x$$

$$\therefore\ u(x,t) = \frac{2}{l}\sum_{r=1}^{\infty}\left\{\left[\int_0^l f(w)\sin\frac{r\pi w}{l}\,\mathrm{d}w\right]\mathrm{e}^{-r^2c^2\pi^2t/l^2}\sin\frac{r\pi x}{l}\right\}$$

Example

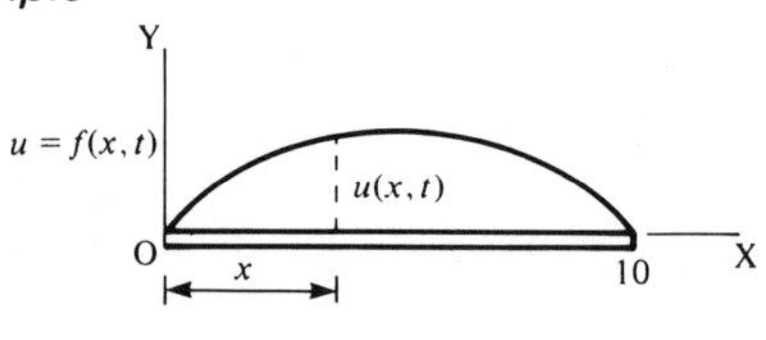

A uniform metal bar, 10 units long, has the end temperatures maintained at 0°C

At $t = 0$, the temperature distribution is defined by $f(x) = x(10-x)$. Find a solution for $u(x,t)$ in the form $u(x,t) = X(x)\,T(t)$ when $c^2 = 4$.

Boundary conditions:

(a) $u(0,t) = 0$
(b) $u(10,t) = 0$ } for all $t \geqslant 0$

(c) $u(x,0) = f(x) = x(10-x)$

$$u = XT \qquad \frac{\partial u}{\partial x} = X'T \qquad \frac{\partial^2 u}{\partial x^2} = X''T \qquad \frac{\partial u}{\partial t} = XT'$$

$$\frac{\partial^2 u}{\partial x^2} = \frac{1}{4}\frac{\partial u}{\partial t} \qquad \therefore \frac{X''}{X} = \frac{1}{4}\frac{T'}{T} = -p^2$$

$$\therefore X'' = -p^2 X \qquad \therefore X'' + p^2 X = 0$$

$$\therefore X = A\cos px + B\sin px$$

$$T' = -4p^2 T \qquad \therefore T = Ce^{-4p^2 t}$$

$$\therefore u(x,t) = \{A\cos px + B\sin px\}\, Ce^{-4p^2 t}$$

i.e. $$\underline{u(x,t) = \{P\cos px + Q\sin px\}\, e^{-4p^2 t}} \qquad (9.17)$$

Using the boundary conditions,

$$u(0,t) = 0 \qquad \therefore 0 = Pe^{-4p^2 t} \qquad \therefore \underline{P = 0}$$

$$\therefore \underline{u(x,t) = Qe^{-4p^2 t}\sin px} \qquad (9.18)$$

Also $$u(10,t) = 0 \qquad \therefore 0 = Qe^{-4p^2 t}\sin 10p$$

$$Q \neq 0 \qquad \therefore \sin 10p = 0 \qquad \therefore 10p = n\pi \qquad \therefore p = \frac{n\pi}{10}$$

$$\therefore \underline{u(x,t) = Qe^{-4p^2 t}\sin\frac{n\pi x}{10}} \qquad (9.19)$$

$$u(x,0) = f(x) = x(10-x) \qquad \therefore x(10-x) = Q\sin\frac{n\pi x}{10}$$

$$Q = 2 \times \text{mean value of } f(x)\sin\frac{n\pi x}{10} \quad \text{between} \quad x = 0 \text{ and } x = 10$$

$$\therefore Q = \frac{2}{10}\int_0^{10} x(10-x)\sin\frac{n\pi x}{10}\,dx$$

$$\therefore 5Q = \left[x(10-x)\left(\frac{-\cos(n\pi x/10)}{n\pi/10}\right)\right]_0^{10} + \frac{10}{n\pi}\int_0^{10}\left(\cos\frac{n\pi x}{10}\right)(10-2x)\,dx$$

$$= \frac{10}{n\pi}\left\{\left[(10-2x)\left(\frac{\sin(n\pi x/10)}{n\pi/10}\right)\right]_0^{10} + \frac{20}{n\pi}\int_0^{10}\sin\frac{n\pi x}{10}\,dx\right\}$$

$$= \frac{10^2}{n^2\pi^2} \times 2\left[\frac{-\cos(n\pi x/10)}{n\pi/10}\right]_0^{10}$$

$$= \frac{2\times 10^3}{n^3\pi^3}\{1-\cos n\pi\} \qquad \begin{cases} n \text{ even}, & \cos n\pi = 1 \\ n \text{ odd}, & \cos n\pi = -1 \end{cases}$$

$$\therefore\ Q = 0 \quad (n \text{ even}); \qquad Q = \frac{4\times 10^3}{5n^3\pi^3} \quad (n \text{ odd})$$

$$\therefore\ u(x,t) = \frac{4\times 10^3}{5n^3\pi^3}\sin\frac{n\pi x}{10}\,\mathrm{e}^{-4n^2\pi^2t/10^2} \qquad n = 1,3,5,\ldots$$

$$\therefore\ u(x,t) = \frac{800}{\pi^3}\left\{\frac{1}{1^3}\sin\frac{\pi x}{10}\,\mathrm{e}^{-0.04\pi^2t} + \frac{1}{3^3}\sin\frac{3\pi x}{10}\,\mathrm{e}^{-0.04\times 3^2\pi^2t}\right.$$

$$\left. + \frac{1}{5^3}\sin\frac{5\pi x}{10}\,\mathrm{e}^{-0.04\times 5^2\pi^2t} + \ldots\right\}$$

9.5 LAPLACE'S EQUATION

This concerns the distribution of a field (temperature, potential, etc.) over a plane area, given certain boundary and initial conditions.

The potential at any point in the plane is defined therefore by the function $u(x,y)$ which is the solution of the two-dimensional Laplace equation

$$\frac{\partial^2 u}{\partial x^2} + \frac{\partial^2 u}{\partial y^2} = 0$$

9.5.1 Solution of the Laplace equation

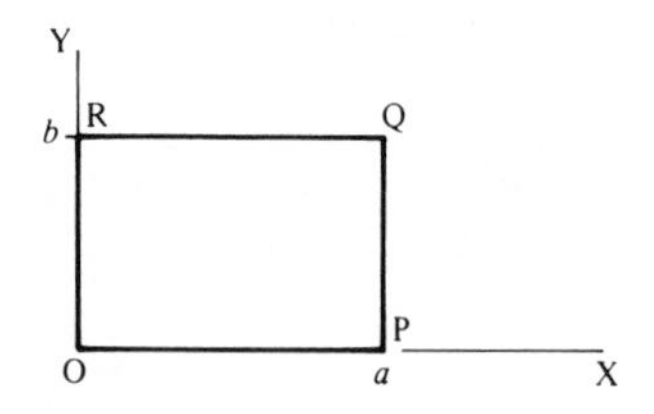

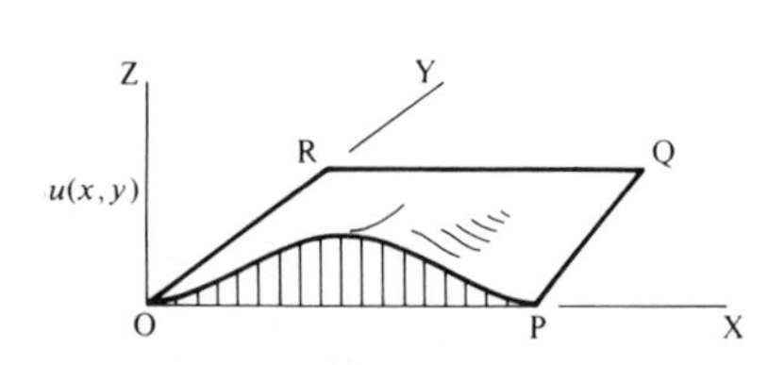

It is required to determine a solution of the Laplace equation for the rectangle bounded by the lines $x = 0, y = 0, x = a, y = b$ which is subject to the following boundary conditions

$u(0,y) = 0 \qquad 0 < y < b$

$u(a,y) = 0 \qquad 0 < y < b$

$u(x,b) = 0 \qquad 0 < x < a$

$u(x,0) = f(x) \qquad 0 < x < a$

The function $u(x,y)$ denotes the potential at any point within the prescribed rectangle.

As usual, let $u(x,y) = X(x)Y(y)$ where X is a function of x only and Y is a function of y only.

$$\frac{\partial u}{\partial x} = X'Y \qquad \frac{\partial^2 u}{\partial x^2} = X''Y \qquad \frac{\partial u}{\partial y} = XY' \qquad \frac{\partial^2 u}{\partial y^2} = XY''$$

$$\frac{\partial^2 u}{\partial x^2} + \frac{\partial^2 u}{\partial y^2} = 0 \qquad \therefore\ X''Y = -XY'' \qquad \therefore\ \frac{Y''}{Y} = -\frac{X''}{X}$$

The left hand side if a function of y only, the right hand side is a function of x only.

Therefore, to be equal, they must each equal the same constant, e.g. p^2.

$$\therefore\ X'' + p^2X = 0 \qquad \therefore\ X = A\cos px + B\sin px$$

$$Y'' - p^2Y = 0 \qquad \therefore\ Y = C\cosh py + D\sinh py = E\sinh p(y+\phi)$$

$$\therefore\ u(x,y) = (A\cos px + B\sin px)E\sinh p(y+\phi)$$

From the boundary conditions,

(a) $$u(0,y) = 0 \qquad \therefore\ 0 = AE\sinh p(y+\phi)$$

$E \neq 0$ or $u(x,y)$ would be identically zero. $\therefore\ \underline{A = 0}$

$$\therefore\ u(x,y) = B\sin px\ E\sinh p(y+\phi) = F\sin px\ \sinh p(y+\phi)$$

(b) $$u(a,y) = 0 \qquad \therefore\ 0 = F\sin pa\ \sinh p(y+\phi)$$

$$F \neq 0 \qquad \therefore\ \sin pa = 0 \qquad \therefore\ pa = n\pi \qquad \therefore\ p = \frac{n\pi}{a} \qquad n = 1, 2, 3, \ldots$$

$$\therefore\ u(x,y) = F\sin\frac{n\pi x}{a}\ \sinh\frac{n\pi}{a}(y+\phi)$$

(c) $$u(x,b) = 0 \qquad \therefore\ 0 = F\sin\frac{n\pi x}{a}\ \sinh\frac{n\pi}{a}(b+\phi)$$

$$\therefore\ \sinh\frac{n\pi}{a}(b+\phi) = 0 \qquad \therefore\ b+\phi = 0 \qquad \therefore\ \phi = -b$$

$$\therefore\ u(x,y) = F\sin\frac{n\pi x}{a}\ \sinh\frac{n\pi}{a}(y-b)$$

Now $$\sinh\frac{n\pi}{a}(y-b) = -\sinh\frac{n\pi}{a}(b-y)$$

$$\therefore\ \underline{u(x,y) = G\sin\frac{n\pi x}{a}\ \sinh\frac{n\pi}{a}(b-y)} \qquad n = 1, 2, 3, \ldots$$

n	$p = \frac{n\pi}{a}$	$u(x,y) = G \sin\frac{n\pi x}{a} \sinh\frac{n\pi}{a}(b-y)$
1	$p_1 = \frac{\pi}{a}$	$u_1 = G_1 \sin\frac{\pi x}{a} \sinh\frac{\pi}{a}(b-y)$
2	$p_2 = \frac{2\pi}{a}$	$u_2 = G_2 \sin\frac{2\pi x}{a} \sinh\frac{2\pi}{a}(b-y)$
3	$p_3 = \frac{3\pi}{a}$	$u_3 = G_3 \sin\frac{3\pi x}{a} \sinh\frac{3\pi}{a}(b-y)$
. . .	. . .	. . .
r	$p_r = \frac{r\pi}{a}$	$u_r = G_r \sin\frac{r\pi x}{a} \sinh\frac{r\pi}{a}(b-y)$
. . .	. . .	. . .

As in the previous examples, a more general solution is therefore given by

$$u = u_1+u_2+u_3+\ldots+u_r+\ldots$$

i.e.

$$u(x,y) = \sum_{r=1}^{\infty} G_r \sin\frac{r\pi x}{a} \sin\frac{r\pi}{a}(b-y)$$

Finally,

$$u(x,0) = f(x)$$

$$\therefore f(x) = \sum_{r=1}^{\infty} G_r \sin\frac{r\pi x}{a} \sinh\frac{r\pi b}{a}$$

Applying Fourier techniques, $G_r \sinh\frac{r\pi b}{a}$ equals twice the mean value of $f(x) \sin\frac{r\pi x}{a}$ over the interval $x = 0$ to $x = a$.

$$\therefore G_r \sinh\frac{r\pi b}{a} = \frac{2}{a}\int_0^a f(x) \sin\frac{r\pi x}{a}\,dx$$

from which the coefficients G_r can be found.

Example

Determine the potential distribution over the rectangle OPQR bounded by the sides $x = 0, y = 0, x = 4, y = 2$ with the boundary conditions stated.

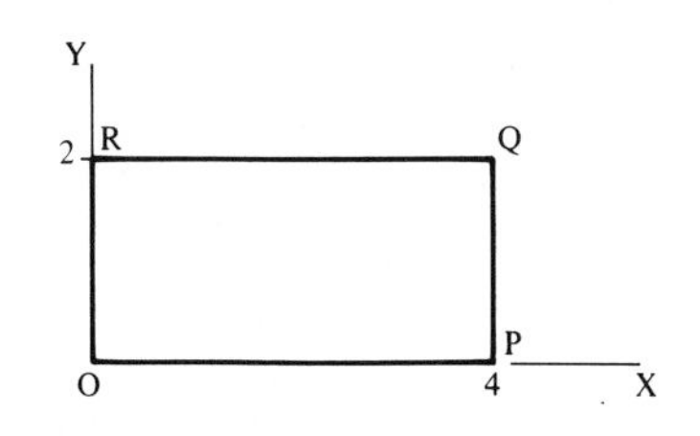

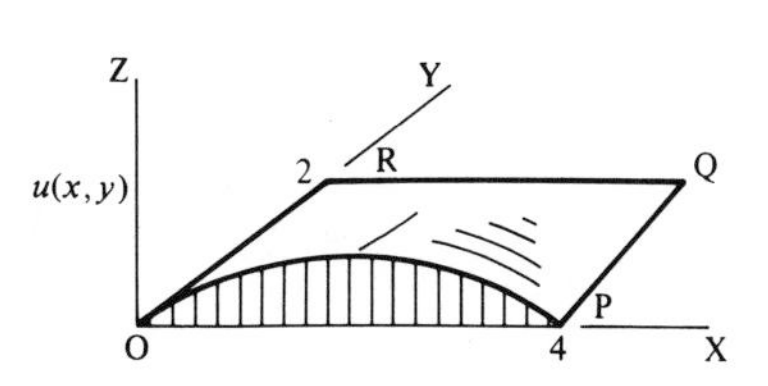

Boundary conditions:

(a) $u(0,y) = 0 \qquad 0<y<2$

(b) $u(4,y) = 0 \qquad 0<y<2$

(c) $u(x,2) = 0 \qquad 0<x<4$

(d) $u(x,0) = x(4-x) \qquad 0<x<4$

$$\frac{\partial^2 u}{\partial x^2} + \frac{\partial^2 u}{\partial y^2} = 0. \qquad \text{Assume} \quad u(x,y) = X(x)Y(y)$$

Then

$$\frac{\partial u}{\partial x} = X'Y \qquad \frac{\partial^2 u}{\partial x^2} = X''Y \qquad \frac{\partial u}{\partial y} = XY' \qquad \frac{\partial^2 u}{\partial y^2} = XY''$$

$$\therefore\ X''Y = -XY'' \qquad \therefore\ \frac{X''}{X} = -\frac{Y''}{Y} \quad (= \text{constant, i.e. } -p^2)$$

$$\therefore\ X'' + p^2X = 0 \qquad \therefore\ X = A\cos px + B\sin px$$

and

$$Y'' - p^2Y = 0 \qquad \therefore\ Y = C\sinh p(y+\phi)$$

$$\therefore\ u(x,y) = (A\cos px + B\sin px)\,C\sinh p(y+\phi)$$

$$= (P\cos px + Q\sin px)\sinh p(y+\phi)$$

From the boundary conditions,

(a) $u(0,y) = 0 \qquad \therefore\ 0 = P\sinh p(y+\phi) \qquad \therefore\ P = 0$

$$\therefore\ u(x,y) = Q\sin px \sinh p(y+\phi)$$

(b) $u(4,y) = 0 \qquad \therefore\ 0 = Q\sin 4p \sinh p(y+\phi)$

$$Q \neq 0 \qquad \therefore\ \sin 4p = 0 \qquad \therefore\ 4p = n\pi \qquad \therefore\ p = \frac{n\pi}{4}$$

$$n = 1, 2, 3, \ldots$$

$$\therefore\ u(x,y) = Q\sin\frac{n\pi x}{4}\sinh\left(\frac{n\pi}{4}(y+\phi)\right)$$

(c) $u(x,2) = 0$ $\qquad \therefore\; 0 = Q\sin\frac{n\pi x}{4}\sinh\left(\frac{n\pi}{4}(2+\phi)\right)$

$$\therefore\; \sinh\frac{n\pi}{4}(2+\phi) = 0 \qquad \therefore\; 2+\phi = 0 \qquad \therefore\; \phi = -2$$

$$\therefore\; u(x,y) = Q\sin\frac{n\pi x}{4}\sinh\left(\frac{n\pi}{4}(y-2)\right)$$

i.e.

$$u(x,y) = R\sin\frac{n\pi x}{4}\sinh\left(\frac{n\pi}{4}(2-y)\right)$$

where $R = -Q$.

n	$p = \frac{n\pi}{4}$	$u(x,y) = R\sin\frac{n\pi x}{4}\sinh\left(\frac{n\pi}{4}(2-y)\right)$
1	$p_1 = \frac{\pi}{4}$	$u_1 = R_1\sin\frac{\pi x}{4}\sinh\left(\frac{\pi}{4}(2-y)\right)$
2	$p_2 = \frac{\pi}{2}$	$u_2 = R_2\sin\frac{\pi x}{2}\sinh\left(\frac{\pi}{2}(2-y)\right)$
3	$p_3 = \frac{3\pi}{4}$	$u_3 = R_3\sin\frac{3\pi x}{4}\sinh\left(\frac{3\pi}{4}(2-y)\right)$
. . .	. . .	. . .
r	$p_r = \frac{r\pi}{4}$	$u_r = R_r\sin\frac{r\pi x}{4}\sinh\left(\frac{r\pi}{4}(2-y)\right)$
. . .	. . .	. . .

$$u = u_1+u_2+u_3+\ldots+u_r+\ldots$$

$$\therefore\; u(x,y) = \sum_{r=1}^{\infty} R_r\sin\frac{r\pi x}{4}\sinh\left(\frac{r\pi}{4}(2-y)\right)$$

Also from the boundary conditions,

(d) $\qquad u(x,0) = f(x) = x(4-x)$

$$\therefore\; f(x) = x(4-x) = \sum_{r=1}^{\infty} R_r\sin\frac{r\pi x}{4}\sinh\frac{r\pi}{2}$$

From the work on Fourier series, $R_r \sinh\dfrac{r\pi}{2}$ is twice the mean value of $f(x)\sin\dfrac{r\pi x}{4}$ over the interval $x = 0$ to $x = 4$.

$$\therefore \; R_r \sinh\frac{r\pi}{2} = \frac{2}{4}\int_0^4 x(4-x)\sin\frac{r\pi x}{4}\,dx$$

$$\therefore \; 2R_r \sinh\frac{r\pi}{2} = \int_0^4 (4x-x^2)\sin\frac{r\pi x}{4}\,dx$$

$$= \left[(4x-x^2)\left(-\frac{4}{r\pi}\cos\frac{r\pi x}{4}\right)\right]_0^4 + \frac{4}{r\pi}\int_0^4 (4-2x)\cos\frac{r\pi x}{4}\,dx$$

$$= \frac{4}{r\pi}\left\{\left[(4-2x)\left(\frac{4}{r\pi}\sin\frac{r\pi x}{4.}\right)\right]_0^4 - \frac{4}{r\pi}\int_0^4 (-2)\sin\frac{r\pi x}{4}\,dx\right\}$$

$$= \frac{4^2}{r^2\pi^2}(2)\left[\frac{-\cos(r\pi x/4)}{r\pi/4}\right]_0^4$$

$$= \frac{2\times 4^3}{r^3\pi^3}(1-\cos r\pi)$$

$$\therefore \; R_r = \frac{4^3}{r^3\pi^3}(1-\cos r\pi)/\sinh\frac{r\pi}{2} \qquad r = 1, 2, 3, \ldots$$

$$\therefore \; u(x,y) = \sum_{r=1}^{\infty}\frac{4^3}{r^3\pi^3}\left(\frac{1-(-1)^r}{\sinh(r\pi/2)}\right)\sinh\left(\frac{r\pi}{2}(2-y)\right)\sin\frac{r\pi x}{4}$$

Exercise 24

1. An elastic string is stretched between two points, 10 units apart. A point on the string, 2 units from its left hand end (the origin), is drawn aside through a distance of 1 unit and released. Apply the one-dimensional wave equation $\dfrac{\partial^2 u}{\partial x^2} = \dfrac{1}{c^2}\dfrac{\partial^2 u}{\partial t^2}$ to determine the displacement at any instant.

2. Solve the boundary value problem $\dfrac{\partial^2 u}{\partial x^2} = \dfrac{1}{c^2}\dfrac{\partial^2 u}{\partial t^2}$ where $c = 3$, subject to the conditions:

$$u(0,t) = 0 \quad\text{and}\quad u(2,t) = 0 \quad \text{for all } t,\; t \geqslant 0$$

$$u(x,0) = x(2-x) \quad\text{and}\quad \left[\frac{\partial u}{\partial t}\right]_{t=0} = 0 \quad\text{for}\quad 0 \leqslant x \leqslant 2$$

3. A stretched string of length 20 cm is set oscillating by displacing its mid-point a distance 1 cm from its rest position and releasing it with zero velocity. State the boundary conditions and solve the one-dimensional wave equation $\frac{\partial^2 u}{\partial x^2} = \frac{1}{c^2}\frac{\partial^2 u}{\partial t^2}$ where $c^2 = 1$ to determine the resulting motion.

4. A perfectly elastic string is stretched between two points 10 cm apart. Its centre point is displaced 2 cm at right-angles to the initial direction of the string and released. Solve the one-dimensional wave equation with $c^2 = 1$ to determine the subsequent motion.

5. A metal bar of length 5 m is insulated along its sides so that heat flow occurs along the bar only. It is initially at a uniform temperature of 20°C and, at $t = 0$, the two ends of the bar are reduced to 0°C and maintained at that temperature. Determine an expression for the temperature at any subsequent time, t seconds, at a distance x metres from one end.

6. One end A of a metal rod AB of length l is kept at 0°C while the other end B is maintained at 50°C until a steady state of temperature is achieved along the rod. At time $t = 0$, the end B is suddenly reduced to 0°C and kept at that temperature. Find an expression for the temperature at any point in the rod, distant x from A, at any time t.

7. The ends of an insulated bar AB, 10 units long, are maintained at 0°C. Initially, the temperature along the bar rises uniformly from each end to 2°C at the mid-point of AB. Determine an expression for the temperature at any point P in the bar distant x from one end at any subsequent time t.

8. Determine the function $u(x, y)$ as a solution of the Laplace equation $\frac{\partial^2 u}{\partial x^2} + \frac{\partial^2 u}{\partial y^2} = 0$ subject to the following boundary conditions:

$$\left.\begin{array}{ll} u = 0 & \text{when } x = 0 \\ u = 0 & \text{when } x = \pi \end{array}\right\} \text{ for all } y \qquad 0 \leqslant y < \infty$$

$$\left.\begin{array}{ll} u \to 0 & \text{when } y \to \infty \\ u = 3 & \text{when } y = 0 \end{array}\right\} \text{ for all } x \qquad 0 \leqslant x \leqslant \pi$$

9. Determine the potential distribution $u(x, y)$ over a square plate bounded by the lines $x = 0, y = 0, x = a, y = a$, subject to the following boundary conditions:

$$\begin{array}{lll} u = 0 & \text{when } x = 0 & 0 \leqslant y \leqslant a \\ u = 0 & \text{when } x = a & 0 \leqslant y \leqslant a \\ u = 0 & \text{when } y = 0 & 0 \leqslant x \leqslant a \\ u = 5 & \text{when } y = a & 0 < x < a \end{array}$$

10. A rectangular plate ABCD is such that AB = CD = 4 units and BC = DA = 2 units. The sides BC, CD, DA are maintained at zero temperature. The fourth side AB has a temperature distribution $f(x) = x(x-4)$ between $x = 0$ and $x = 4$. Determine an expression for the steady-state temperature at any point in the plate.

9.6 REVISION SUMMARY

1. Ordinary second order linear differential equations

(a) Equations of the form $a\dfrac{d^2y}{dx^2} + b\dfrac{dy}{dx} + cy = 0$

Auxiliary equation: $am^2 + bm + c = 0$

(i) Real and different roots: $m = m_1$ and $m = m_2$

Solution: $y = Ae^{m_1x} + Be^{m_2x}$

(ii) Real and equal roots: $m = m_1$ (twice)

Solution: $y = e^{m_1x}(A + Bx)$

(iii) Complex roots: $m = \alpha \pm j\beta$

Solution: $y = e^{\alpha x}(A\cos\beta x + B\sin\beta x)$

(b) Equations of the form $\dfrac{d^2y}{dx^2} \pm n^2y = 0$

(i) $\dfrac{d^2y}{dx^2} + n^2y = 0$ $\qquad y = A\cos nx + B\sin nx$

(ii) $\dfrac{d^2y}{dx^2} - n^2y = 0$ $\qquad y = A\cosh nx + B\sinh nx$

or $y = Ae^{nx} + Be^{-nx}$

or $y = A\sinh n(x + \phi)$

(c) Equations of the form $a\dfrac{d^2y}{dx^2} + b\dfrac{dy}{dx} + cy = f(x)$

General solution = complementary function + particular integral

2. Partial differential equations

$$u = f(x, y, t, \ldots)$$

Linear equations: If $u = u_1$, $u = u_2$, $u = u_3$, ..., are solutions, so also is

$$u = u_1 + u_2 + u_3 + \ldots, \text{ i.e. } u = \sum_{r=1}^{\infty} u_r.$$

(a) *Wave equation* – transverse vibrations of an elastic string

$$\frac{\partial^2 u}{\partial x^2} = \frac{1}{c^2}\frac{\partial^2 u}{\partial t^2} \qquad c^2 = \frac{T}{\rho}$$

where T = tension of string

ρ = mass per unit length of string

(b) *Heat conduction equation* – heat flow in a uniform finite bar

$$\frac{\partial^2 u}{\partial x^2} = \frac{1}{c^2}\frac{\partial u}{\partial t} \qquad c^2 = \frac{k}{\sigma\rho}$$

where k = thermal conductivity of material

σ = specific heat of the material

ρ = mass per unit length of bar

(c) *Laplace equation* – distribution of a field over a plane area

$$\frac{\partial^2 u}{\partial x^2} + \frac{\partial^2 u}{\partial y^2} = 0$$

3. Separating the variables

Assume $$u(x,y) = X(x)Y(y)$$

Then $\dfrac{\partial u}{\partial x} = X'Y;$ $\quad \dfrac{\partial^2 u}{\partial x^2} = X''Y;$ $\quad \dfrac{\partial u}{\partial y} = XY';$ $\quad \dfrac{\partial^2 u}{\partial y^2} = XY''$

Substitute in the given partial differential equation and form separate differential equations to give $X(x)$ and $Y(y)$. Determine arbitrary functions by application of initial and boundary conditions.

Chapter 10

FOURIER INTEGRALS AND TRANSFORMS

10.1 EXTENDED PERIOD

If we consider a periodic function $f(x)$ of period $2L$ defined by

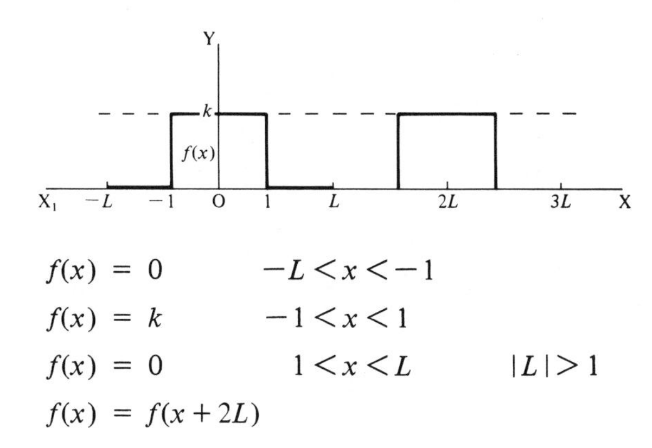

$$\begin{aligned} f(x) &= 0 & -L < x < -1 & \\ f(x) &= k & -1 < x < 1 & \\ f(x) &= 0 & 1 < x < L & \quad |L| > 1 \\ f(x) &= f(x+2L) & & \end{aligned}$$

then the function can be represented by a series of the form

$$f(x) = \tfrac{1}{2}a_0 + \sum_{n=1}^{\infty}\left\{a_n \cos\frac{n\pi x}{L} + b_n \sin\frac{n\pi x}{L}\right\}$$

where
$$a_0 = \frac{1}{L}\int_{-L}^{L} f(x)\,dx$$

$$a_n = \frac{1}{L}\int_{-L}^{L} f(x)\cos\frac{n\pi x}{L}\,dx$$

$$b_n = \frac{1}{L}\int_{-L}^{L} f(x)\sin\frac{n\pi x}{L}\,dx \qquad n = 1, 2, 3, \ldots$$

The usual analysis gives the results

$$a_0 = \frac{2k}{L}; \qquad a_n = \frac{2k}{n\pi}\sin\frac{n\pi}{L}; \qquad b_n = 0$$

all of which are numerical values depending on the values of the constants k, L and n.

If the original definition of $f(x)$ remains unchanged, but the value of L is increased indefinitely, then the function will become that of a single pulse of amplitude k and pulse-width 2 units.

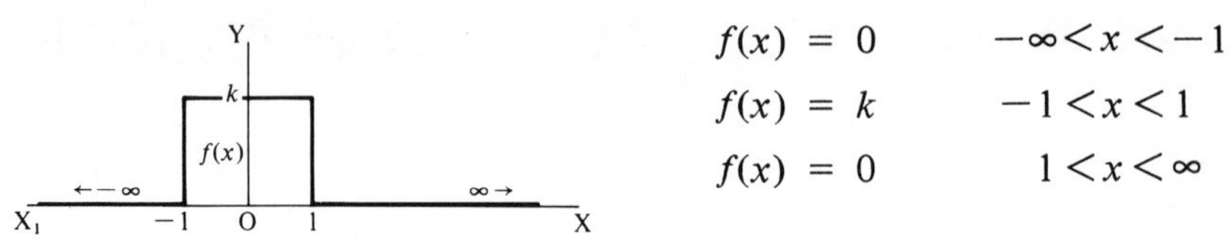

$$\begin{aligned} f(x) &= 0 \qquad -\infty < x < -1 \\ f(x) &= k \qquad -1 < x < 1 \\ f(x) &= 0 \qquad 1 < x < \infty \end{aligned}$$

The function will then cease to be periodic and the normal Fourier series quoted above will no longer apply. The function can now, however, be represented in the form of a *Fourier integral.*

10.2 FOURIER INTEGRAL

As before, consider a periodic function $f(x)$ of period $2L$ defined over the interval $-L$ to L. Then

$$f(x) = \tfrac{1}{2}a_0 + \sum_{n=1}^{\infty}\left\{a_n \cos\frac{n\pi x}{L} + b_n \sin\frac{n\pi x}{L}\right\} \tag{10.1}$$

where

$$a_0 = \frac{1}{L}\int_{-L}^{L} f(u)\,du$$

$$a_n = \frac{1}{L}\int_{-L}^{L} f(u)\cos\frac{n\pi u}{L}\,du$$

$$b_n = \frac{1}{L}\int_{-L}^{L} f(u)\sin\frac{n\pi u}{L}\,du$$

Note that the variable x is replaced by u in evaluating the definite integrals to distinguish it from the independent variable x of the series in (10.1).

Substituting the expressions for a_0, a_n and b_n in (10.1) gives

$$\begin{aligned} f(x) &= \frac{1}{2L}\int_{-L}^{L} f(u)\,du + \frac{1}{L}\sum_{n=1}^{\infty}\left\{\left[\int_{-L}^{L} f(u)\cos\frac{n\pi u}{L}\,du\right]\cos\frac{n\pi x}{L}\right. \\ &\qquad\qquad \left. + \left[\int_{-L}^{L} f(u)\sin\frac{n\pi u}{L}\,du\right]\sin\frac{n\pi x}{L}\right\} \\ &= \frac{1}{2L}\int_{-L}^{L} f(u)\,du + \frac{1}{L}\sum_{n=1}^{\infty}\left\{\int_{-L}^{L} f(u)\left[\cos\frac{n\pi x}{L}\cos\frac{n\pi u}{L}\right.\right. \\ &\qquad\qquad \left.\left. + \sin\frac{n\pi x}{L}\sin\frac{n\pi u}{L}\right]du\right\} \end{aligned}$$

But $$\cos A \cos B + \sin A \sin B = \cos (A-B)$$

$$\therefore\ f(x) = \frac{1}{2L}\int_{-L}^{L} f(u)\,du + \frac{1}{L}\sum_{n=1}^{\infty}\left\{\int_{-L}^{L} f(u)\cos\frac{n\pi}{L}(x-u)\,du\right\}$$

Since $\cos\left[-\frac{n\pi}{L}(x-u)\right] = \cos\left[\frac{n\pi}{L}(x-u)\right]$, this can be written

$$f(x) = \frac{1}{2L}\sum_{n=-\infty}^{\infty}\left\{\int_{-L}^{L} f(u)\cos\frac{n\pi}{L}(x-u)\,du\right\} \tag{10.2}$$

Note that when $n = 0$, the relevant term in (10.2) becomes $\frac{1}{2L}\int_{-L}^{L} f(u)\,du$, i.e. $\frac{1}{2}a_0$.

If L increases, but remains finite, then $\frac{\pi}{L}$ decreases. Therefore, we can write $\frac{\pi}{L} = \Delta\lambda$. Then

$$f(x) = \sum_{n=-\infty}^{\infty}\frac{\Delta\lambda}{2\pi}\left\{\int_{-L}^{L} f(u)\cos n\Delta\lambda(x-u)\,du\right\}$$

If $n \to \infty$ while $\Delta\lambda \to 0$ in such a way that $n\Delta\lambda \to \infty$, then it can be shown that, subject to certain conditions, $n\Delta\lambda$ can be replaced by a continuous variable, say λ. Then, as $L \to \infty$, (10.2) can be written as a double integral

$$f(x) = \frac{1}{2\pi}\int_{-\infty}^{\infty}\int_{-L}^{L} f(u)\cos\lambda(x-u)\,du\,d\lambda$$

i.e. $$\underline{f(x) = \frac{1}{\pi}\int_{0}^{\infty}\int_{-\infty}^{\infty} f(u)\cos\lambda(x-u)\,du\,d\lambda}$$

This is one form of the Fourier integral expansion of $f(x)$. The expansion is valid if

(a) $f(x)$ and $f'(x)$ are piecewise continuous in the range $-\infty$ to ∞

(b) $\int_{-\infty}^{\infty} f(x)\,dx$ converges in the range $-\infty$ to ∞.

As is the case of the Fourier series,

(c) the result holds good at a point of continuity of $f(x)$

(d) at a point of discontinuity, $f(x)$ is replaced by $\frac{f(x+0)+f(x-0)}{2}$.

Example

Express in Fourier integral form, the function $f(x)$ such that

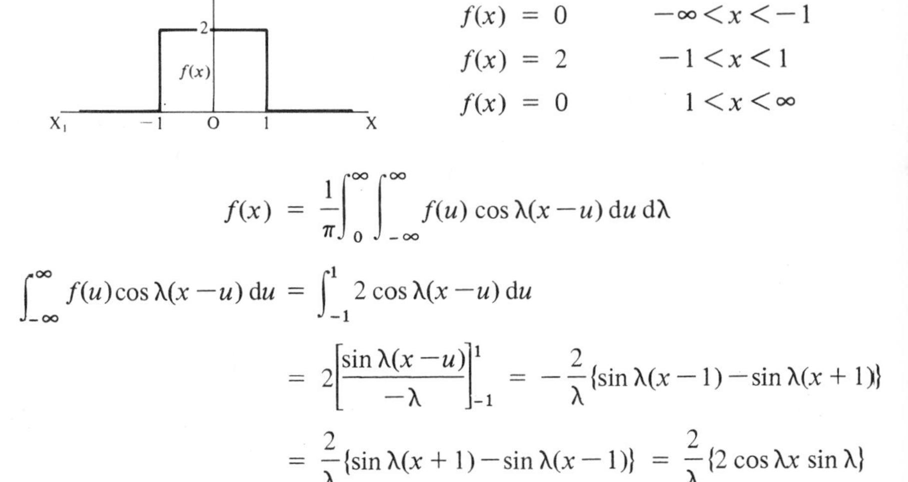

$$f(x) = 0 \qquad -\infty < x < -1$$
$$f(x) = 2 \qquad -1 < x < 1$$
$$f(x) = 0 \qquad 1 < x < \infty$$

$$f(x) = \frac{1}{\pi}\int_0^\infty \int_{-\infty}^\infty f(u) \cos \lambda(x-u)\, du\, d\lambda$$

$$\int_{-\infty}^\infty f(u) \cos \lambda(x-u)\, du = \int_{-1}^1 2 \cos \lambda(x-u)\, du$$

$$= 2\left[\frac{\sin \lambda(x-u)}{-\lambda}\right]_{-1}^1 = -\frac{2}{\lambda}\{\sin \lambda(x-1) - \sin \lambda(x+1)\}$$

$$= \frac{2}{\lambda}\{\sin \lambda(x+1) - \sin \lambda(x-1)\} = \frac{2}{\lambda}\{2 \cos \lambda x \sin \lambda\}$$

$$\therefore\ f(x) = \frac{1}{\pi}\int_0^\infty \frac{2}{\lambda}\{2 \cos \lambda x \sin \lambda\}\, d\lambda$$

$$\therefore\ f(x) = \frac{4}{\pi}\int_0^\infty \frac{1}{\lambda}(\cos \lambda x \sin \lambda)\, d\lambda$$

10.3 OTHER FORMS OF THE FOURIER INTEGRAL

The Fourier integral representation of a function $f(x)$ can be stated in various forms.

(a) $f(x) = \int_0^\infty \{A(\lambda) \cos \lambda x + B(\lambda) \sin \lambda x\}\, d\lambda$ (Form I)

where
$$A(\lambda) = \frac{1}{\pi}\int_{-\infty}^\infty f(u) \cos \lambda u\, du$$

$$B(\lambda) = \frac{1}{\pi}\int_{-\infty}^\infty f(u) \sin \lambda u\, du$$

(b) $f(x) = \frac{1}{\pi}\int_0^\infty \int_{-\infty}^\infty f(u) \cos \lambda(x-u)\, du\, d\lambda$ (Form II)

(c) $f(x) = \frac{1}{2\pi}\int_{-\infty}^{\infty}\int_{-\infty}^{\infty} f(u)\, e^{j\lambda(x-u)}\, du\, d\lambda$ (Form III)

(d) $f(x) = \frac{1}{2\pi}\int_{-\infty}^{\infty} e^{j\lambda x}\int_{-\infty}^{\infty} f(u)\, e^{-j\lambda u}\, du\, d\lambda$ (Form IV)

It can readily be established that the four forms are equivalent to each other, as shown in the examples below.

Example 1

$$\text{Form II} \equiv f(x) = \frac{1}{\pi}\int_{0}^{\infty}\int_{-\infty}^{\infty} f(u)\cos\lambda(x-u)\, du\, d\lambda$$

$$= \frac{1}{\pi}\int_{0}^{\infty}\left\{\int_{-\infty}^{\infty} f(u)\cos\lambda(x-u)\, du\right\} d\lambda$$

$$= \frac{1}{\pi}\int_{0}^{\infty}\int_{-\infty}^{\infty} f(u)\{\cos\lambda x\cos\lambda u + \sin\lambda x\sin\lambda u\} du\, d\lambda$$

$$= \int_{0}^{\infty}\{A(\lambda)\cos\lambda x + B(\lambda)\sin\lambda x\}\, d\lambda$$

where

$$A(\lambda) = \frac{1}{\pi}\int_{-\infty}^{\infty} f(u)\cos\lambda u\, du$$

$$B(\lambda) = \frac{1}{\pi}\int_{-\infty}^{\infty} f(u)\sin\lambda u\, du$$

i.e. $\underline{\text{Form II} \equiv \text{Form I}}$

Example 2

$$\text{Form II} \equiv f(x) = \frac{1}{\pi}\int_{0}^{\infty}\int_{-\infty}^{\infty} f(u)\cos\lambda(x-u)\, du\, d\lambda$$

Since $\cos\lambda(x-u)$ is an even function in λ,

$$f(x) = \frac{1}{2\pi}\int_{-\infty}^{\infty}\int_{-\infty}^{\infty} f(u)\cos\lambda(x-u)\, du\, d\lambda \quad \text{(i)}$$

Also, $\sin\lambda(x-u)$ is an odd function in λ

$$\therefore\ 0 = \frac{1}{2\pi}\int_{-\infty}^{\infty}\int_{-\infty}^{\infty} f(u)\sin\lambda(x-u)\, du\, d\lambda \quad \text{(ii)}$$

$$\therefore \text{ (i)} + \text{j(ii)} = f(x) = \frac{1}{2\pi}\int_{-\infty}^{\infty}\int_{-\infty}^{\infty} f(u)\{\cos\lambda(x-u) + \text{j}\sin\lambda(x-u)\}\,du\,d\lambda$$

$$\therefore f(x) = \frac{1}{2\pi}\int_{-\infty}^{\infty}\int_{-\infty}^{\infty} f(u)\,e^{j\lambda(x-u)}\,du\,d\lambda$$

i.e. $\underline{\text{Form II} \equiv \text{Form III}}$

Example 3

$$\text{Form IV} \equiv f(x) = \frac{1}{2\pi}\int_{-\infty}^{\infty} e^{j\lambda x}\int_{-\infty}^{\infty} f(u)\,e^{-j\lambda u}\,du\,d\lambda$$

$$= \frac{1}{2\pi}\int_{-\infty}^{\infty}\int_{-\infty}^{\infty} e^{j\lambda x} f(u)\,e^{-j\lambda u}\,du\,d\lambda$$

$$= \frac{1}{2\pi}\int_{-\infty}^{\infty}\int_{-\infty}^{\infty} f(u)\,e^{j\lambda(x-u)}\,du\,d\lambda$$

i.e. $\underline{\text{Form III} \equiv \text{Form IV}}$

10.3.1 Special cases of the Fourier integral

1. *If $f(x)$ is an odd function*

$$f(x) = \int_0^{\infty}\{A(\lambda)\cos\lambda x + B(\lambda)\sin\lambda x\}\,d\lambda$$

Odd function $\quad \therefore f(-x) = -f(x) \quad \therefore A(\lambda) = 0$

$$\therefore f(x) = \int_0^{\infty} B(\lambda)\sin\lambda x\,d\lambda$$

where

$$B(\lambda) = \frac{1}{\pi}\int_{-\infty}^{\infty} f(u)\sin\lambda u\,du$$

$$\therefore f(x) = \frac{1}{\pi}\int_0^{\infty}\left\{\int_{-\infty}^{\infty} f(u)\sin\lambda u\,du\right\}\sin\lambda x\,d\lambda$$

$$= \frac{1}{\pi}\int_0^{\infty}\sin\lambda x\int_{-\infty}^{\infty} f(u)\sin\lambda u\,du\,d\lambda$$

$$\underline{f(x) = \frac{2}{\pi}\int_0^{\infty}\sin\lambda x\int_0^{\infty} f(u)\sin\lambda u\,du\,d\lambda} \qquad \text{(Form V)}$$

This is the *Fourier sine integral.*

2. *If $f(x)$ is an even function*

In this case, $f(-x) = f(x)$ $\therefore\ B(\lambda) = 0$

$$\therefore\ f(x) = \int_0^\infty A(\lambda) \cos \lambda x \, d\lambda$$

where
$$A(\lambda) = \frac{1}{\pi}\int_{-\infty}^{\infty} f(u) \cos \lambda u \, du$$

$$\therefore\ f(x) = \frac{1}{\pi}\int_0^\infty \cos \lambda x \int_{-\infty}^{\infty} f(u) \cos \lambda u \, du \, d\lambda$$

$$\therefore\ f(x) = \frac{2}{\pi}\int_0^\infty \cos \lambda x \int_0^\infty f(u) \cos \lambda u \, du \, d\lambda \qquad \text{(Form VI)}$$

This is the *Fourier cosine integral.*

10.3.2 Summary

Reviewing the six results, we have

I $$f(x) = \int_0^\infty \{A(\lambda) \cos \lambda x + B(\lambda) \sin \lambda x\} \, d\lambda$$

where
$$A(\lambda) = \frac{1}{\pi}\int_{-\infty}^{\infty} f(u) \cos \lambda u \, du$$

$$B(\lambda) = \frac{1}{\pi}\int_{-\infty}^{\infty} f(u) \sin \lambda u \, du$$

II $$f(x) = \frac{1}{\pi}\int_0^\infty \int_{-\infty}^{\infty} f(u) \cos \lambda(x-u) \, du \, d\lambda$$

III $$f(x) = \frac{1}{2\pi}\int_{-\infty}^{\infty} \int_{-\infty}^{\infty} f(u) \, e^{j\lambda(x-u)} \, du \, d\lambda$$

IV $$f(x) = \frac{1}{2\pi}\int_{-\infty}^{\infty} e^{j\lambda x} \int_{-\infty}^{\infty} f(u) \, e^{-j\lambda u} \, du \, d\lambda$$

V $f(x)$ an *odd function* :

$$f(x) = \frac{2}{\pi}\int_0^\infty \sin \lambda x \int_0^\infty f(u) \sin \lambda u \, du \, d\lambda$$

VI $f(x)$ an *even function* :

$$f(x) = \frac{2}{\pi}\int_0^\infty \cos \lambda x \int_0^\infty f(u) \cos \lambda u \, du \, d\lambda$$

Example 1

Express in Fourier integral form the function $f(x)$ defined by

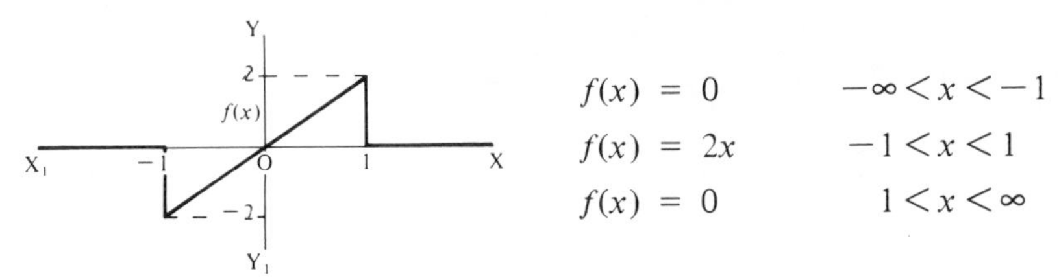

$$f(x) = 0 \qquad -\infty < x < -1$$
$$f(x) = 2x \qquad -1 < x < 1$$
$$f(x) = 0 \qquad 1 < x < \infty$$

$f(x)$ is an odd function. We therefore use the form

$$f(x) = \frac{2}{\pi}\int_0^\infty \sin \lambda x \int_0^\infty f(u) \sin \lambda u \, du \, d\lambda$$

$$\int_0^\infty f(u) \sin \lambda u \, du = \int_0^1 2u \sin \lambda u \, du$$

$$= 2\left\{\left[u\left(\frac{-\cos \lambda u}{\lambda}\right)\right]_0^1 + \frac{1}{\lambda}\int_0^1 \cos \lambda u \, du\right\}$$

$$= 2\left\{\frac{-\cos \lambda}{\lambda} + \frac{1}{\lambda}\left[\frac{\sin \lambda u}{\lambda}\right]_0^1\right\}$$

$$= 2\left\{\frac{-\cos \lambda}{\lambda} + \frac{1}{\lambda^2}\sin \lambda\right\}$$

$$\therefore \; f(x) = \frac{4}{\pi}\int_0^\infty \sin \lambda x \left\{\frac{\sin \lambda}{\lambda^2} - \frac{\cos \lambda}{\lambda}\right\} d\lambda$$

Example 2

A function $f(x)$ is defined as follows

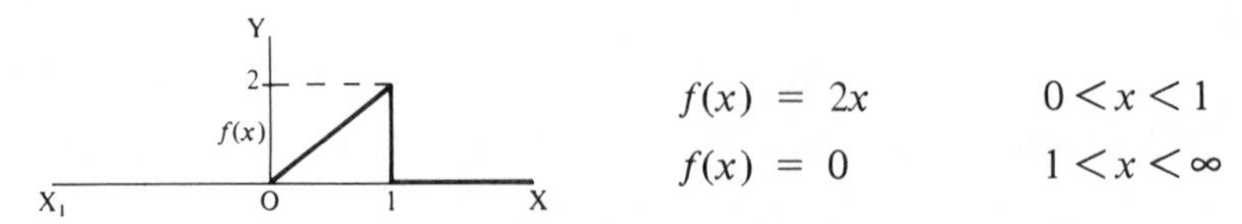

$$f(x) = 2x \qquad 0 < x < 1$$
$$f(x) = 0 \qquad 1 < x < \infty$$

Note that, in this case, the function is not defined for negative values of x. As with the study of half-range Fourier series, we can therefore consider the waveform extended to the left as an even function or as an odd function and then express the function either as a Fourier cosine integral or as a Fourier sine integral. The result obtained will, of course, apply to $f(x)$ only for values of $x \geqslant 0$.

Let us, therefore, choose to regard the function as part of an even function and express it as a Fourier cosine integral.

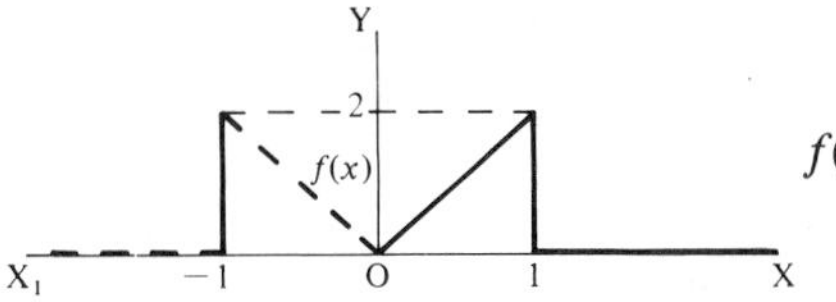

$$f(x) = \frac{2}{\pi}\int_0^\infty \cos\lambda x \, d\lambda \int_0^\infty f(u)\cos\lambda u \, du$$

$$\int_0^\infty f(u)\cos\lambda u\,du = \int_0^1 2u\cos\lambda u\,du$$

$$= 2\left\{\left[u\left(\frac{\sin\lambda u}{\lambda}\right)\right]_0^1 - \frac{1}{\lambda}\int_0^1 \sin\lambda u\,du\right\}$$

$$= 2\left\{\frac{\sin\lambda}{\lambda} - \frac{1}{\lambda}\left[\frac{-\cos\lambda u}{\lambda}\right]_0^1\right\}$$

$$= 2\left\{\frac{\sin\lambda}{\lambda} + \frac{\cos\lambda - 1}{\lambda^2}\right\}$$

$$\therefore\ f(x) = \frac{2}{\pi}\int_0^\infty \cos\lambda x\ 2\left\{\frac{\sin\lambda}{\lambda} + \frac{\cos\lambda - 1}{\lambda^2}\right\} d\lambda$$

$$\therefore\ f(x) = \frac{4}{\pi}\int_0^\infty \cos\lambda x \left\{\frac{\sin\lambda}{\lambda} + \frac{\cos\lambda - 1}{\lambda^2}\right\} d\lambda$$

Exercise 25

1. A function $f(x)$ is defined by

$$f(x) = 0 \qquad -\infty < x < 0$$
$$f(x) = 5 \qquad 0 < x < 1$$
$$f(x) = 0 \qquad 1 < x < \infty$$

Express the function as a Fourier integral.

2. A function $f(x)$ is defined in the range $x = 0$ to $x = \infty$ by:

$$f(x) = x - 2 \qquad 0 < x < 2$$
$$f(x) = 0 \qquad 2 < x < \infty$$

Express the function as a Fourier cosine integral.

3. A function $f(x)$ is defined by

$$f(x) = 2x \qquad 0 < x < 1$$
$$f(x) = 3 - x \qquad 1 < x < 3$$
$$f(x) = 0 \qquad 3 < x < \infty$$

Determine the Fourier sine integral to represent the function and state for what range of values of x the integral is valid.

4. Express as a Fourier integral, the function defined by

$$f(x) = 0 \qquad -\infty < x < 0$$
$$f(x) = x^2 \qquad 0 < x < 2$$
$$f(x) = 0 \qquad 2 < x < \infty$$

5. A function $f(x)$ is defined as follows

$$f(x) = 0 \qquad -\infty < x < -2$$
$$f(x) = x + 2 \qquad -2 < x < -1$$
$$f(x) = 1 \qquad -1 < x < 1$$
$$f(x) = 2 - x \qquad 1 < x < 2$$
$$f(x) = 0 \qquad 2 < x < \infty$$

Represent the function in terms of a Fourier integral.

10.4 AMPLITUDE AND PHASE

One form of the Fourier integral that we have used is

$$f(x) = \int_0^\infty \{A(\lambda)\cos\lambda x + B(\lambda)\sin\lambda x\}\,d\lambda$$

where
$$A(\lambda) = \frac{1}{\pi}\int_{-\infty}^{\infty} f(u)\cos\lambda u\,du$$

$$B(\lambda) = \frac{1}{\pi}\int_{-\infty}^{\infty} f(u)\sin\lambda u\,du$$

This can be written in amplitude–phase form

$$f(x) = \int_0^\infty R(\lambda)\cos(\lambda x + \phi)\,d\lambda$$

where
$$R(\lambda) = \sqrt{\{A(\lambda)\}^2 + \{B(\lambda)\}^2}$$

$$\tan\phi = -\frac{B(\lambda)}{A(\lambda)}$$

The Fourier integral representation therefore contains information giving both the amplitude $R(\lambda)$ and the phase ϕ of $f(x)$.

Example

(a) Determine the Fourier integral representation of $f(x)$ where

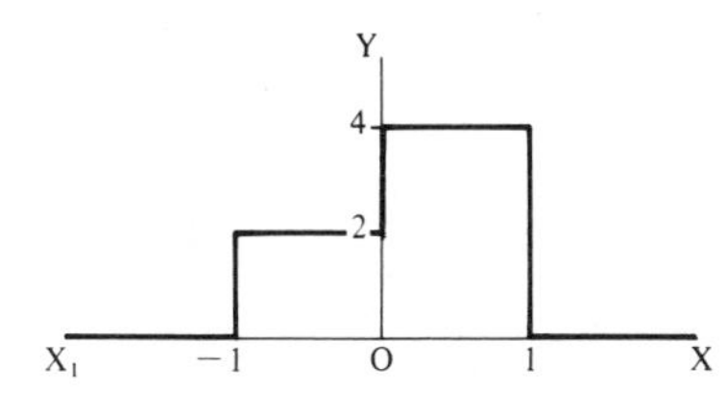

$$\begin{aligned} f(x) &= 0 & -\infty < x < -1 \\ f(x) &= 2 & -1 < x < 0 \\ f(x) &= 4 & 0 < x < 1 \\ f(x) &= 0 & 1 < x < \infty \end{aligned}$$

(b) Obtain an expression for the amplitude $R(\lambda)$.

(c) Plot the spectrum of $f(x)$ between $\lambda = 0$ and $\lambda = 10$.

(a)
$$f(x) = \int_0^\infty \{A(\lambda)\cos\lambda x + B(\lambda)\sin\lambda x\}\,d\lambda$$

where
$$A(\lambda) = \frac{1}{\pi}\int_{-\infty}^{\infty} f(u)\cos\lambda u\,du$$

$$B(\lambda) = \frac{1}{\pi}\int_{-\infty}^{\infty} f(u)\sin\lambda u\,du$$

$$\therefore\ \pi A(\lambda) = \int_{-1}^{0} 2\cos\lambda u\,du + \int_0^1 4\cos\lambda u\,du$$

$$= 2\left[\frac{\sin\lambda u}{\lambda}\right]_{-1}^{0} + 4\left[\frac{\sin\lambda u}{\lambda}\right]_0^1$$

$$= \frac{6\sin\lambda}{\lambda} \qquad \therefore\ A(\lambda) = \frac{1}{\pi}\frac{6\sin\lambda}{\lambda}$$

$$\pi B(\lambda) = \int_{-1}^{0} 2\sin\lambda u\,du + \int_0^1 4\sin\lambda u\,du$$

$$= 2\left[\frac{-\cos\lambda u}{\lambda}\right]_{-1}^{0} + 4\left[\frac{-\cos\lambda u}{\lambda}\right]_0^1$$

$$= \frac{2}{\lambda} - \frac{2\cos\lambda}{\lambda} \qquad \therefore\ B(\lambda) = \frac{1}{\pi}\frac{2(1-\cos\lambda)}{\lambda}$$

$$\therefore\ f(x) = \frac{1}{\pi}\int_0^\infty \left\{\frac{6\sin\lambda\cos\lambda x}{\lambda} + \frac{2(1-\cos\lambda)\sin\lambda x}{\lambda}\right\}d\lambda$$

(b)
$$\{R(\lambda)\}^2 = \{A(\lambda)\}^2 + \{B(\lambda)\}^2$$
$$= \frac{36\sin^2\lambda}{\pi^2\lambda^2} + \frac{4(1-\cos\lambda)^2}{\pi^2\lambda^2}$$

which simplifies to

$$= \frac{8(5-\cos\lambda-4\cos^2\lambda)}{\pi^2\lambda^2}$$
$$= \frac{8(5+4\cos\lambda)(1-\cos\lambda)}{\pi^2\lambda^2}$$
$$\therefore \text{ Amplitude} = R(\lambda) = \frac{\sqrt{8(5+4\cos\lambda)(1-\cos\lambda)}}{\pi\lambda}$$

(c) Plotting values of $R(\lambda)$ against values of λ between $\lambda = 0$ and $\lambda = 10$ gives the required spectrum of $f(x)$.

λ	1	2	3	4	5	6	7	8	9	10
$R(\lambda)$	1.634	0.978	0.432	0.447	0.378	0.089	0.181	0.253	0.161	0.157

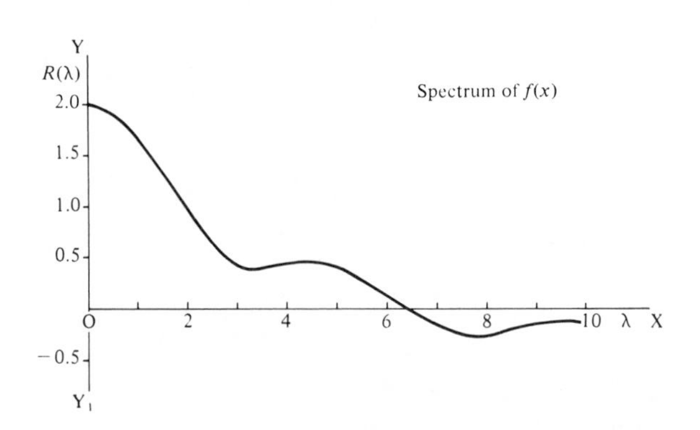

Note:

(a) Since $f(x)$ is defined for all values of x, λ can have all values and the spectrum formed by plotting $R(\lambda)$ against λ is therefore a *continuous spectrum.* This contrasts with the spectrum of a periodic function which was seen earlier to consist of a set of *discrete ordinates.*

(b) The amplitude of a sinusoidal component is normally defined as being positive, but, in this context, it is customary, when appropriate, to draw the spectrum as in the diagram shown.

Exercise 26

For each of the following functions $f(x)$, determine an expression $R(\lambda)$ for the amplitude and hence plot the spectrum of $f(x)$, i.e. the graph of $R(\lambda)$ against λ, for the range $\lambda = 0$ to $\lambda = 10$.

1. $f(x) = 0 \quad -\infty < x < 0$
   $f(x) = 5 \quad 0 < x < 2$
   $f(x) = 0 \quad 2 < x < \infty$

2. $f(x) = 0 \quad -\infty < x < 0$
   $f(x) = 2x \quad 0 < x < 1$
   $f(x) = 0 \quad 1 < x < \infty$

3. $f(x) = 0 \quad -\infty < x < -2$
   $f(x) = -1 \quad -2 < x < 0$
   $f(x) = 4 \quad 0 < x < 2$
   $f(x) = 0 \quad 2 < x < \infty$

4. $f(x) = 0 \quad -\infty < x < 0$
   $f(x) = e^{-x} \sin x \quad 0 < x < \infty$

0.5 FOURIER TRANSFORMS

10.5.1 Fourier complex transform

One form of the Fourier integral, previously established, was

$$f(x) = \frac{1}{2\pi}\int_{-\infty}^{\infty} e^{j\lambda x} \int_{-\infty}^{\infty} f(u)\, e^{-j\lambda u}\, du\, d\lambda \tag{10.3}$$

If we write
$$F(\lambda) = \int_{-\infty}^{\infty} f(u)\, e^{-j\lambda u}\, du$$

then
$$f(x) = \frac{1}{2\pi}\int_{-\infty}^{\infty} F(\lambda)\, e^{j\lambda x}\, d\lambda$$

$F(\lambda)$ is then called the *Fourier transform* (or *Fourier complex transform*) of $f(x)$ and is written $F(\lambda) = \mathscr{F}\{f(x)\}$.

$$\therefore\ \mathscr{F}\{f(x)\} = F(\lambda) = \int_{-\infty}^{\infty} f(u)\, e^{-j\lambda u}\, du \tag{10.4}$$

In reverse, if $F(\lambda)$ is known, then $f(x)$ is called the *inverse Fourier transform* of $F(\lambda)$ and this is written $f(x) = \mathscr{F}^{-1}\{F(\lambda)\}$

i.e.
$$\mathscr{F}^{-1}\{F(\lambda)\} = f(x) = \frac{1}{2\pi}\int_{-\infty}^{\infty} F(\lambda)\, e^{j\lambda x}\, d\lambda \tag{10.5}$$

Example

Find the Fourier transform of the function $f(x)$ defined by

$f(x) = 0 \quad -\infty < x < -2$
$f(x) = 1 \quad -2 < x < 2$
$f(x) = 0 \quad 2 < x < \infty$

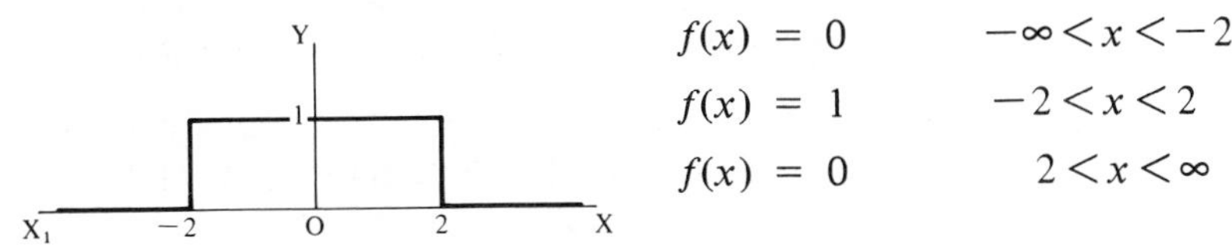

From the definition of the Fourier transform

$$F(\lambda) = \int_{-\infty}^{\infty} f(u)\,\mathrm{e}^{-\mathrm{j}\lambda u}\,du = \int_{-2}^{2} 1\mathrm{e}^{-\mathrm{j}\lambda u}\,du$$

$$= \left[\frac{\mathrm{e}^{-\mathrm{j}\lambda u}}{-\mathrm{j}\lambda}\right]_{-2}^{2} = \frac{\mathrm{e}^{\mathrm{j}\lambda 2} - \mathrm{e}^{-\mathrm{j}\lambda 2}}{\mathrm{j}\lambda}$$

$$= \frac{2}{\lambda}\left\{\frac{\mathrm{e}^{\mathrm{j}\lambda 2} - \mathrm{e}^{-\mathrm{j}\lambda 2}}{\mathrm{j}2}\right\} = \frac{2}{\lambda}\sin 2\lambda \qquad \lambda \neq 0$$

$$\therefore\ F(\lambda) = \frac{2\sin 2\lambda}{\lambda}$$

10.5.2 Fourier sine transform

If $f(x)$ is an odd function, i.e. $f(-x) = -f(x)$, the Fourier integral can be written in the form

$$f(x) = \frac{2}{\pi}\int_{0}^{\infty} \sin\lambda x \int_{0}^{\infty} f(u)\sin\lambda u\,du\,d\lambda$$

Let $$F_s(\lambda) = \int_{0}^{\infty} f(u)\sin\lambda u\,du$$

Then $$f(x) = \frac{2}{\pi}\int_{0}^{\infty} F_s(\lambda)\sin\lambda x\,d\lambda$$

$F_s(\lambda) = \int_0^{\infty} f(u)\sin\lambda u\,du$ is called the *Fourier sine transform* of $f(x)$ and, in reverse, $f(x)$ is the *inverse Fourier sine transform* of $F_s(\lambda)$,

i.e. $$F_s(\lambda) = \mathscr{F}_s\{f(x)\}$$

and $$f(x) = \mathscr{F}_s^{-1}\{F_s(\lambda)\}$$

10.5.3 Fourier cosine transform

If $f(x)$ is an even function, i.e. $f(-x) = f(x)$, the Fourier integral can be written in the form

$$f(x) = \frac{2}{\pi}\int_{0}^{\infty} \cos\lambda x \int_{0}^{\infty} f(u)\cos\lambda u\,du\,d\lambda$$

Let $$F_c(\lambda) = \int_0^\infty f(u)\cos\lambda u\,du$$

Then $$f(x) = \frac{2}{\pi}\int_0^\infty F_c(\lambda)\cos\lambda x\,d\lambda$$

$F_c(\lambda) = \int_0^\infty f(u)\cos\lambda u\,du$ is called the *Fourier cosine transform* of $f(x)$,

i.e. $$F_c(\lambda) = \mathscr{F}_c\{f(x)\}$$

and $$f(x) = \mathscr{F}_c^{-1}\{F_c(\lambda)\}$$

10.5.4 Summary

(a) *Fourier complex transform*

$$F(\lambda) = \int_{-\infty}^{\infty} f(u)\,e^{-j\lambda u}\,du \qquad f(x) = \frac{1}{2\pi}\int_{-\infty}^{\infty} F(\lambda)\,e^{j\lambda x}\,d\lambda$$

(b) *Fourier sine transform* for $f(x)$ an odd function

$$F_s(\lambda) = \int_0^\infty f(u)\sin\lambda u\,du \qquad f(x) = \frac{2}{\pi}\int_0^\infty F_s(\lambda)\sin\lambda x\,d\lambda$$

(c) *Fourier cosine transform* for $f(x)$ an even function

$$F_c(\lambda) = \int_0^\infty f(u)\cos\lambda u\,du \qquad f(x) = \frac{2}{\pi}\int_0^\infty F_c(\lambda)\cos\lambda x\,d\lambda$$

Example

Consider the function defined by

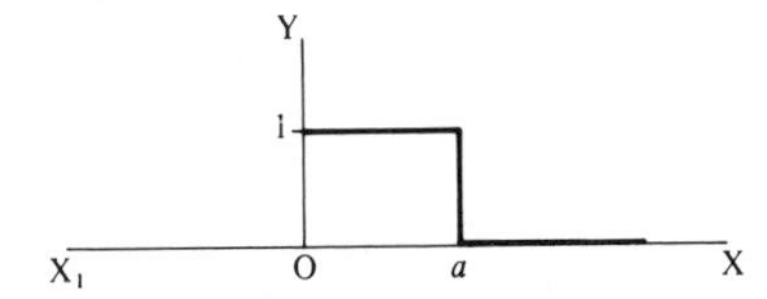

$$f(x) = 1 \qquad 0 < x < a$$
$$f(x) = 0 \qquad a < x < \infty$$

If we regard this as half of an even function, we can determine its Fourier cosine transform.

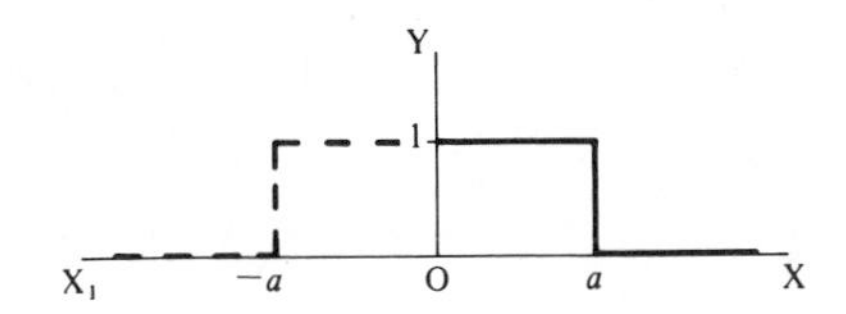

$$F_c(\lambda) = \int_0^\infty f(u)\cos\lambda u\,du = \int_0^a 1\cos\lambda u\,du$$

$$= \left[\frac{\sin\lambda u}{\lambda}\right]_0^a = \frac{1}{\lambda}\{\sin\lambda a\} \qquad \therefore\ F_c(\lambda) = \frac{\sin\lambda a}{\lambda} \qquad (10.6)$$

Alternatively, if we regard the function as half of an odd function, we can determine its sine transform.

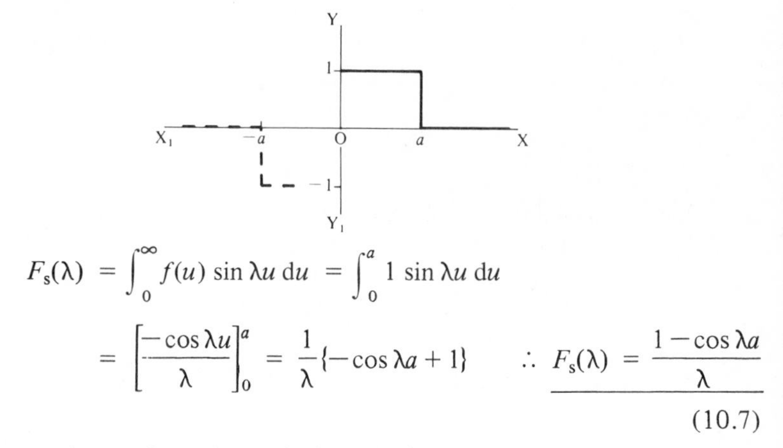

$$F_s(\lambda) = \int_0^\infty f(u)\sin\lambda u\,du = \int_0^a 1\sin\lambda u\,du$$

$$= \left[\frac{-\cos\lambda u}{\lambda}\right]_0^a = \frac{1}{\lambda}\{-\cos\lambda a + 1\} \qquad \therefore\ F_s(\lambda) = \frac{1-\cos\lambda a}{\lambda} \qquad (10.7)$$

In each case, the result applies only for $x > 0$.

Inverse transforms

From (10.6)

$$f(x) = \frac{2}{\pi}\int_0^\infty F_c(\lambda)\cos\lambda x\,d\lambda$$

$$\therefore\ f(x) = \frac{2}{\pi}\int_0^\infty \frac{\sin\lambda a\cos\lambda x}{\lambda}\,d\lambda$$

From (10.7)

$$f(x) = \frac{2}{\pi}\int_0^\infty F_s(\lambda)\sin\lambda x\,d\lambda$$

$$\therefore\ f(x) = \frac{2}{\pi}\int_0^\infty \frac{(1-\cos\lambda a)\sin\lambda x}{\lambda}\,d\lambda$$

Exercise 27

1. Form the Fourier complex transform of the function $f(x)$ defined by

$$f(x) = 0 \qquad -\infty < x < -1$$
$$f(x) = 1-x^2 \qquad -1 < x < 1$$
$$f(x) = 0 \qquad 1 < x < \infty$$

2. Determine the Fourier sine transform of $f(x)$ where

$$f(x) = 0 \qquad -\infty < x < -1$$
$$f(x) = x \qquad -1 < x < 1$$
$$f(x) = 0 \qquad 1 < x < \infty$$

3. A function $f(x)$ is defined by

$$f(x) = 2x \qquad 0 < x < 1$$
$$f(x) = 2 \qquad 1 < x < 3$$
$$f(x) = 0 \qquad 3 < x < \infty$$

   Determine the Fourier cosine transform of $f(x)$.

4. Determine the Fourier sine transform of the function $f(x) = e^{-mx}$ defined in the domain $x = 0$ to $x = \infty$.

5. Find the Fourier transform of the function $f(x)$ defined by

$$f(x) = 0 \qquad -\infty < x < -\frac{\pi}{2}$$
$$f(x) = \sin 2x \qquad -\frac{\pi}{2} < x < \frac{\pi}{2}$$
$$f(x) = 0 \qquad \frac{\pi}{2} < x < \infty$$

10.6 FOURIER INTEGRALS IN THE SOLUTION OF BOUNDARY VALUE PROBLEMS

Fourier integral techniques can be applied in the solution of boundary value problems to establish the arbitrary constants or functions involved. The following two examples give instances where this is possible.

Example 1

Solve the equation

$$\frac{\partial^2 v}{\partial x^2} = \frac{1}{4}\frac{\partial v}{\partial y}$$

subject to the boundary conditions

$$v(0, y) = 0 \qquad \text{for all} \quad y \geqslant 0$$
$$v(k, y) = 0 \qquad \text{for all} \quad y \geqslant 0$$
$$v(x, 0) = f(x) \qquad \text{where} \quad f(x) = x(k-x) \quad \text{for} \quad 0 < x < k$$

Assume a solution of the form $v(x,y) = X(x)Y(y)$.

Then $\quad \dfrac{\partial v}{\partial x} = X'Y; \qquad \dfrac{\partial^2 v}{\partial x^2} = X''Y; \qquad \dfrac{\partial v}{\partial y} = XY'$

$$\therefore\ X''Y = \frac{1}{4}XY' \qquad \text{i.e.}\quad \frac{X''}{X} = \frac{1}{4}\frac{Y'}{Y} \qquad (= -p^2 \text{ say})$$

$$\therefore\ X'' + p^2X = 0 \qquad \therefore\ X = A\cos px + B\sin px$$

and $\quad \dfrac{Y'}{Y} = -4p^2 \qquad \therefore\ \ln Y = -4p^2y \qquad \therefore\ Y = Ce^{-4p^2y}$

$$\therefore\ v = (A\cos px + B\sin px)\,Ce^{-4p^2y}$$

i.e. $\quad v = (P\cos px + Q\sin px)\,e^{-4p^2y}$

From the boundary conditions,

(a) $\quad v(0,y) = 0 \qquad \therefore\ 0 = Pe^{-4p^2y} \qquad \therefore\ P = 0$

$$\therefore\ v = Q\sin px\ e^{-4p^2y}$$

(b) $\quad v(k,y) = 0 \qquad \therefore\ 0 = Q\sin pk\ e^{-4p^2y}$

$Q \neq 0$ or $v(x,y)$ would be identically zero.

$$\therefore\ \sin pk = 0 \qquad \therefore\ pk = r\pi \qquad \therefore\ p = \frac{r\pi}{k} \qquad r = 1, 2, 3, \ldots$$

$$\therefore\ v_r = Q_r \sin\frac{r\pi x}{k}\, e^{-4r^2\pi^2y/k^2} \qquad r = 1, 2, 3, \ldots$$

$\therefore$ General solution is

$$v(x,y) = \sum_{r=1}^{\infty} Q_r \sin\frac{r\pi x}{k}\, e^{-4r^2\pi^2y/k^2}$$

This must also satisfy the remaining boundary condition

(c) $\quad v(x,0) = f(x) \qquad$ where $\qquad f(x) = x(k-x)\quad$ for $\quad 0<x<k$

$$\therefore\ x(k-x) = \sum_{r=1}^{\infty} Q_r \sin\frac{r\pi x}{k}$$

$$\therefore\ Q_r = \frac{2}{k}\int_0^k f(u)\sin pu\, du \qquad \text{where } p = \frac{r\pi}{k}$$

$$= \frac{2}{k}\int_0^k u(k-u)\sin pu\, du$$

Integrating by parts, this simplifies to

$$Q_r = \frac{4}{kp^3}(1-\cos pk) = \frac{4k^2}{r^3\pi^3}(1-\cos r\pi)$$

$$\therefore \quad Q_r = \frac{4k^2}{r^3\pi^3}[1-(-1)^r]$$

$$\therefore \quad v(x,y) = \frac{4k^2}{\pi^3}\sum_{r=1}^{\infty}\left\{\left[\frac{1-(-1)^r}{r^3}\right]\sin\frac{r\pi x}{k}\,e^{-4r^2\pi^2 y/k^2}\right\}$$

Example 2

Solve the equation $\dfrac{\partial^2 v}{\partial x^2}+\dfrac{\partial^2 v}{\partial y^2} = 0$ subject to the following boundary conditions

(a) $\dfrac{\partial v}{\partial x} = 0$ when $x = 0$ for all $y \geqslant 0$

(b) $\dfrac{\partial v}{\partial y} = 0$ when $y = 0$ for all $x \geqslant 0$

(c) $v(x,y) = f(x)$ when $y = 1$ where $f(x) = (1-x) \quad 0<x<1$

$f(x) = 0 \quad 1<x<\infty$

As usual, assume a solution of the form $v(x,y) = X(x)Y(y)$ i.e. $v = XY$

$$\therefore \quad \frac{\partial v}{\partial x} = X'Y; \qquad \frac{\partial^2 v}{\partial x^2} = X''Y; \qquad \frac{\partial v}{\partial y} = XY'; \qquad \frac{\partial^2 v}{\partial y^2} = XY''$$

Then $\quad X''Y + XY'' = 0 \quad$ i.e. $\quad X''Y = -XY''$

$$\therefore \quad \frac{X''}{X} = -\frac{Y''}{Y} \qquad (= -p^2 \text{ say})$$

$$\therefore \quad X'' + p^2X = 0 \qquad \therefore \quad X = A\cos px + B\sin px$$

and

$$Y'' - p^2Y = 0 \qquad \therefore \quad Y = C\cosh py + D\sinh py$$

$$\therefore \quad v(x,y) = (A\cos px + B\sin px)(C\cosh py + D\sinh py)$$

$$\frac{\partial v}{\partial x} = (-Ap\sin px + Bp\cos px)(C\cosh py + D\sinh py)$$

From the boundary condition (a), $\dfrac{\partial v}{\partial x} = 0$ when $x = 0$.

$$\therefore \quad 0 = Bp(C\cosh py + D\sinh py) \qquad \therefore \quad \underline{B = 0}$$

$$\therefore \quad v(x,y) = A\cos px(C\cosh py + D\sinh py)$$

Also $$\frac{\partial v}{\partial y} = A\cos px(Cp\sinh py + Dp\cosh py)$$

and from the boundary condition (b), $\dfrac{\partial v}{\partial y} = 0$ when $y = 0$.

$$\therefore\ 0 = A\cos px(Dp) \qquad \therefore\ \underline{D = 0}$$

$$\therefore\ v(x,y) = A\cos px\, C\cosh py = P\cos px\cosh py,$$

where $P = AC$.

Similar solutions will depend on the values given to p. A more general solution is therefore given by

$$v(x,y) = \sum_{r=1}^{\infty} P_r \cos p_r x \cosh p_r y$$

But p can have any positive value. Therefore the result can be written

$$v(x,y) = \int_0^\infty A(\lambda)\cos\lambda x\cosh\lambda y\, d\lambda$$

From the final boundary condition (c), when $y = 1$

$$v(x,y) = f(x) = 1-x \qquad 0<x<1$$

$$f(x) = 0 \qquad 1<x<\infty$$

$$\therefore\ 1-x = \int_0^\infty A(\lambda)\cos\lambda x\cosh\lambda\, d\lambda$$

Using the Fourier cosine integral

$$A(\lambda)\cosh\lambda = \frac{2}{\pi}\int_0^\infty f(u)\cos\lambda u\, du$$

$$= \frac{2}{\pi}\int_0^1 (1-u)\cos\lambda u\, du$$

Integrating by parts, this simplifies to

$$A(\lambda)\cosh\lambda = \frac{2}{\pi}\frac{(1-\cos\lambda)}{\lambda^2}$$

$$\therefore\ A(\lambda) = \frac{2}{\pi\cosh\lambda}\left(\frac{1-\cos\lambda}{\lambda^2}\right)$$

Finally, therefore

$$v(x,y) = \int_0^\infty A(\lambda)\cos\lambda x\cosh\lambda y\, d\lambda$$

$$= \int_0^\infty \frac{2}{\pi \cosh \lambda}\left(\frac{1-\cos\lambda}{\lambda^2}\right)\cos\lambda x \cosh\lambda y \, d\lambda$$

$$\therefore \; v(x,y) = \frac{2}{\pi}\int_0^\infty \frac{(1-\cos\lambda)}{\lambda^2 \cosh\lambda}\cos\lambda x \cosh\lambda y \, d\lambda$$

Exercise 28

1. Assuming a solution of the equation $\dfrac{\partial^2 v}{\partial x^2} = \dfrac{1}{c^2}\dfrac{\partial v}{\partial t}$ in the form $v(x,t) = X(x)T(t)$, solve the equation, given that at $t = 0$, $v = f(x) = e^{-x}$.

2. Solve the equation $\dfrac{\partial^2 v}{\partial x^2} = \dfrac{\partial v}{\partial t}$ subject to the boundary conditions

   (a) $v = 0$ when $x = 0$ for all $t \geqslant 0$

   (b) $v = f(x)$ when $t = 1$ for all $x \geqslant 0$.

3. Solve the Laplace equation $\dfrac{\partial^2 v}{\partial x^2} + \dfrac{\partial^2 v}{\partial y^2} = 0$ given that at $y = 0$,

$$v(x,y) = f(x) = 1 + x \qquad 0 < x < 1$$
$$f(x) = 0 \qquad 1 < x < \infty$$

4. Determine the solution of the equation $\dfrac{\partial^2 v}{\partial x^2} = \dfrac{1}{c^2}\dfrac{\partial^2 v}{\partial t^2}$ subject to the following boundary conditions

   (a) $v = 0$ when $x = 0$ } for all $t \geqslant 0$

   (b) $v = 0$ when $x = 2$ }

   (c) $v = 2x$ when $t = 0$ } for $0 \leqslant x \leqslant 2$

   (d) $\dfrac{\partial v}{\partial t} = x^2$ when $t = 0$ }

5. Solve $\dfrac{\partial^2 v}{\partial x^2} + \dfrac{\partial^2 v}{\partial y^2} = 0$ given that

   (a) $v = 0$ when $x = 0$ for all $y \geqslant 0$

   (b) $\dfrac{\partial v}{\partial y} = 0$ when $y = 0$ for all $x \geqslant 0$

   (c) $u = f(x)$ when $y = 2$, where $f(x) = x \qquad 0 < x < 1$

   $f(x) = 0 \qquad 1 < x < \infty$

10.7 REVISION SUMMARY

1. Fourier integral

Alternative forms:

(a) $$f(x) = \int_0^\infty \{A(\lambda) \cos \lambda x + B(\lambda) \sin \lambda x\}\, d\lambda \qquad \text{(Form I)}$$

where $$A(\lambda) = \frac{1}{\pi}\int_{-\infty}^{\infty} f(u) \cos \lambda u\, du$$

$$B(\lambda) = \frac{1}{\pi}\int_{-\infty}^{\infty} f(u) \sin \lambda u\, du$$

(b) $$f(x) = \frac{1}{\pi}\int_0^\infty \int_{-\infty}^{\infty} f(u) \cos \lambda(x-u)\, du\, d\lambda \qquad \text{(Form II)}$$

(c) $$f(x) = \frac{1}{2\pi}\int_{-\infty}^{\infty} \int_{-\infty}^{\infty} f(u)\, e^{j\lambda(x-u)}\, du\, d\lambda \qquad \text{(Form III)}$$

(d) $$f(x) = \frac{1}{2\pi}\int_{-\infty}^{\infty} e^{j\lambda x} \int_{-\infty}^{\infty} f(u)\, e^{-j\lambda u}\, du\, d\lambda \qquad \text{(Form IV)}$$

(e) *Fourier sine integral* for $f(x)$ an odd function

$$f(x) = \frac{2}{\pi}\int_0^\infty \sin \lambda x \int_0^\infty f(u) \sin \lambda u\, du\, d\lambda \qquad \text{(Form V)}$$

(f) *Fourier cosine integral* for $f(x)$ an even function

$$f(x) = \frac{2}{\pi}\int_0^\infty \cos \lambda x \int_0^\infty f(u) \cos \lambda u\, du\, d\lambda \qquad \text{(Form VI)}$$

2. Amplitude and phase

$$f(x) = \int_0^\infty \{A(\lambda) \cos \lambda x + B(\lambda) \sin \lambda x\}\, d\lambda$$

$$\text{Amplitude} = R(\lambda) = \sqrt{\{A(\lambda)\}^2 + \{B(\lambda)\}^2}$$

$$\text{Phase} = \phi = \arctan\left\{-\frac{B(\lambda)}{A(\lambda)}\right\}$$

3. Spectrum

Plot of amplitude against frequency i.e. $R(\lambda)$ against λ.

4. Fourier transforms

(a) *Fourier complex transform*

$$F(\lambda) = \int_{-\infty}^{\infty} f(u)\, e^{-j\lambda u}\, du$$

Inverse transform

$$f(x) = \frac{1}{2\pi}\int_{-\infty}^{\infty} F(\lambda)\, e^{j\lambda x}\, d\lambda$$

(b) *Fourier sine transform*

$$F_s(\lambda) = \int_{0}^{\infty} f(u) \sin \lambda u\, du$$

$$f(x) = \frac{2}{\pi}\int_{0}^{\infty} F_s(\lambda) \sin \lambda x\, d\lambda$$

(c) *Fourier cosine transform*

$$F_c(\lambda) = \int_{0}^{\infty} f(u) \cos \lambda u\, du$$

$$f(x) = \frac{2}{\pi}\int_{0}^{\infty} F_c(\lambda) \cos \lambda x\, d\lambda$$

ANSWERS

EXERCISE 1 (p. 2)

1.

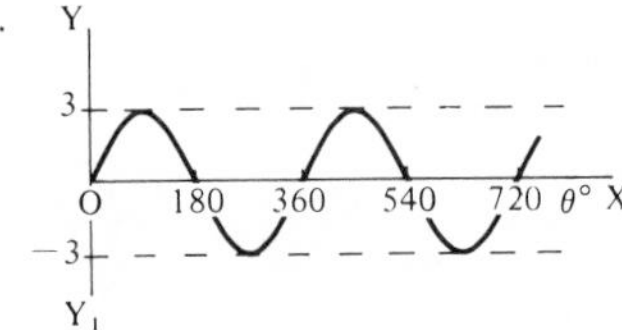

2.

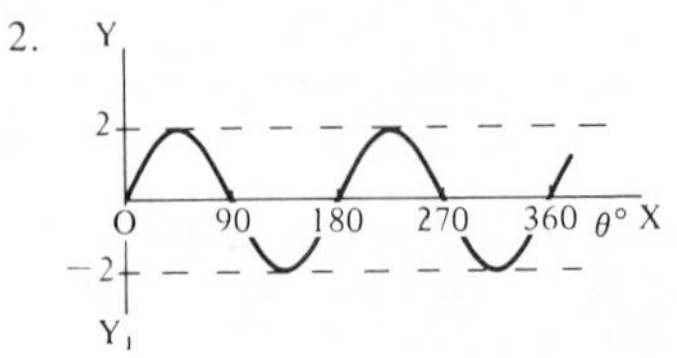

3.

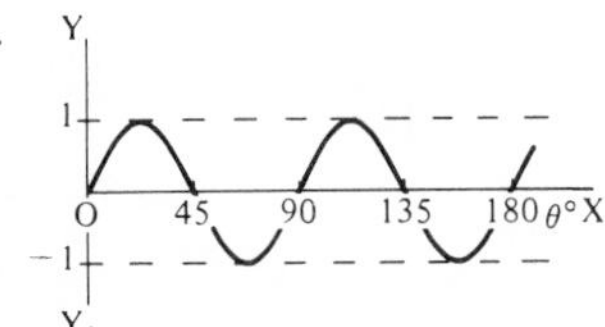

4.

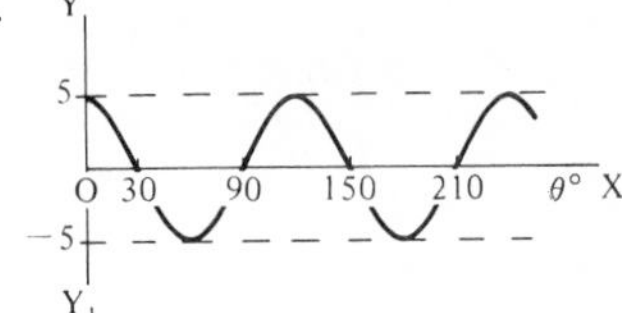

5.

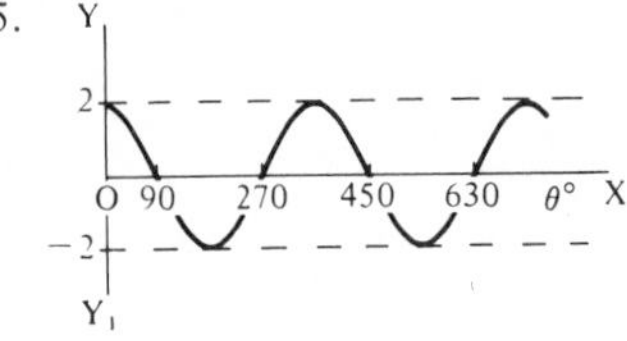

6.

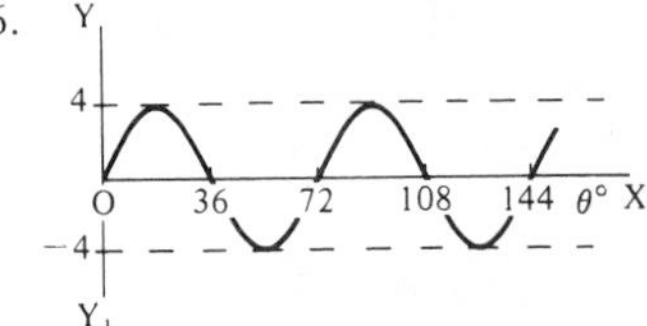

7.

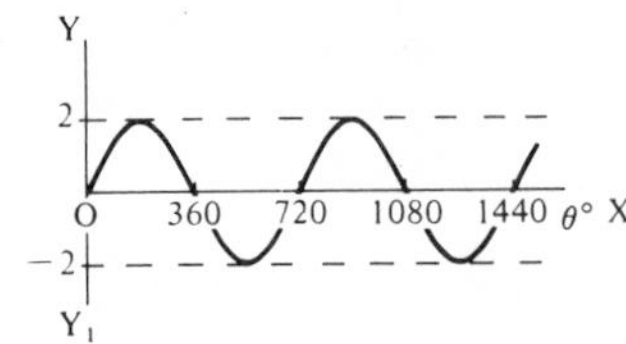

8.

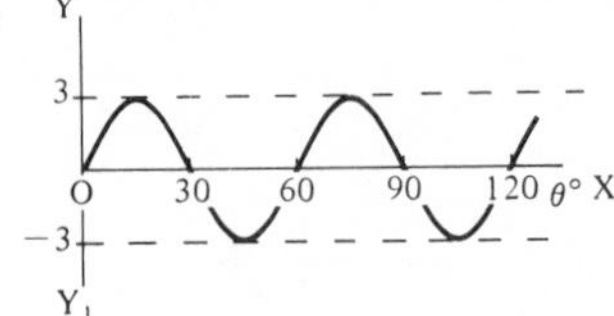

9.

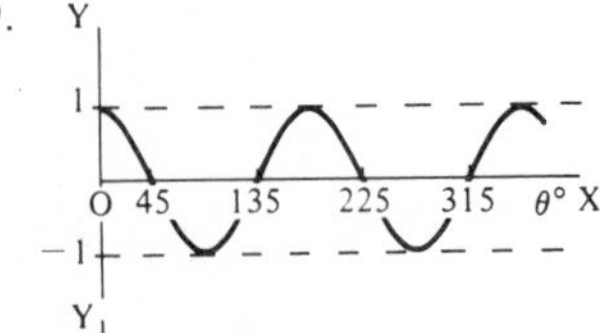

10.

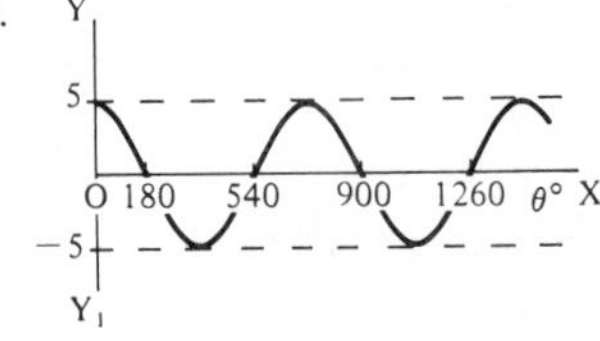

EXERCISE 2 (p. 6)

1. 180°	2. 72°	3. 720°	4. 90°	5. 120°
6. 600°	7. 360°	8. 360°	9. 360°	10. 180°

EXERCISE 3 (p. 7)

1. 8 ms	2. 6 ms	3. 9 ms	4. 11 ms	5. 10 ms

EXERCISE 4 (p. 10)

1. $f(x) = 5 \qquad 0 < x < 6$

   $f(x) = -2 \qquad 6 < x < 10$

   $f(x) = f(x + 10)$

2. $f(x) = 6 - \dfrac{3x}{4} \qquad 0 < x < 8$

   $f(x) = 0 \qquad 8 < x < 12$

   $f(x) = f(x + 12)$

3. $f(x) = \dfrac{4x}{3} \qquad 0 < x < 3$

   $f(x) = 4 \qquad 3 < x < 7$

   $f(x) = f(x + 7)$

4. $f(x) = 8 - \dfrac{4x}{3} \qquad 0 < x < 9$

   $f(x) = f(x + 9)$

5. $f(x) = 8 \qquad 0 < x < 4$

   $f(x) = 5 \qquad 4 < x < 7$

   $f(x) = 0 \qquad 7 < x < 10$

   $f(x) = f(x + 10)$

6. $f(x) = \dfrac{4x}{5} \qquad 0 < x < 5$

   $f(x) = 8 - \dfrac{4x}{5} \qquad 5 < x < 10$

   $f(x) = 0 \qquad 10 < x < 12$

   $f(x) = f(x + 12)$

EXERCISE 5 (p. 12)

1.

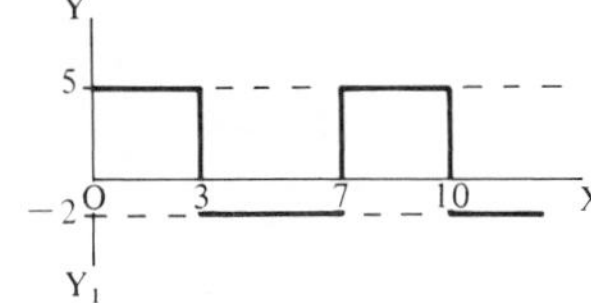

2.

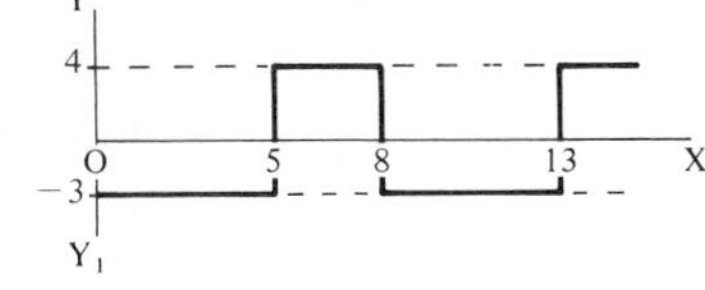

3.

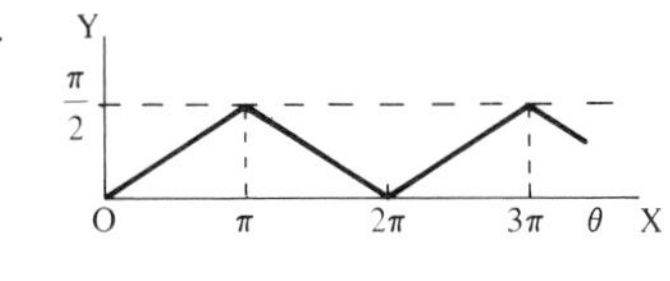

4.

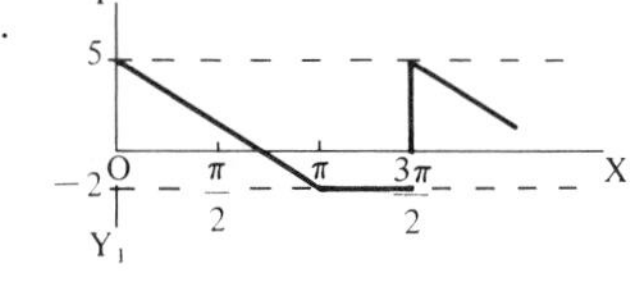

5.

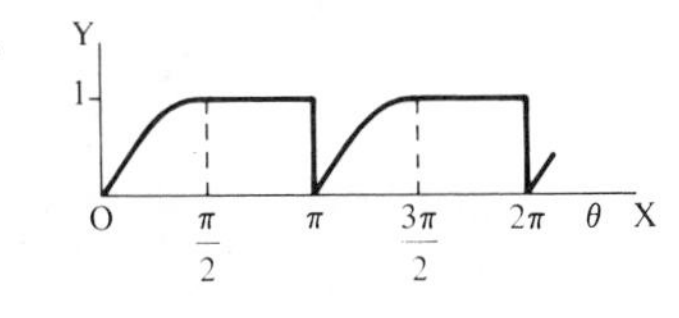

EXERCISE 6 (p. 16)

1. $5.831 \sin(\theta + 59°2')$
2. $7.211 \sin(\theta - 33°41')$
3. $3.606 \cos(\theta - 56°19')$
4. $5.385 \cos(\theta + 21°48')$
5. $5.000 \sin(2\theta - 36°52')$
6. $2.953 \cos(5\theta + 28°18')$
7. $3.996 \sin(\theta + 31°42')$
8. $4.950 \cos(3\theta - 20°42')$

EXERCISE 7 (p. 22)

1. $f(x) = \dfrac{e^{3x}}{3}\left\{x^2 - \dfrac{2x}{3} + \dfrac{2}{9}\right\} + C$
2. $f(x) = 3x^2\left\{\ln x - \dfrac{1}{2}\right\} + C$
3. $f(x) = \dfrac{e^{2x}}{5}\{2 \sin x - \cos x\} + C$
4. $f(x) = \dfrac{1}{4}\{(2x^2 - 1) \sin 2x + 2x \cos 2x\} + C$

5. $f(x) = -\frac{e^{-5x}}{5}\left\{x + \frac{1}{5}\right\} + C$

6. $f(t) = \frac{1}{\omega}\left\{\frac{\sin \omega t}{\omega} - t \cos \omega t\right\} + C$

7. $\frac{2\pi}{3}$ 8. $\frac{\pi}{8}$ 9. -0.2121 10. 0.3248

EXERCISE 8 (p. 25)

1. Yes
2. Yes
3. No – infinite discontinuity at $x = 0$
4. Yes
5. No – infinite discontinuities at $x = \pm\frac{\pi}{2}$
6. No – two valued
7. Yes
8. No – infinite discontinuities

EXERCISE 9 (p. 37)

1. $f(x) = 2 + \frac{4}{\pi}\left\{\sin x - \frac{1}{2}\sin 2x + \frac{1}{3}\sin 3x - \ldots\right\}$

2. $f(x) = \frac{5}{2} + \frac{10}{\pi}\left\{\sin x + \frac{1}{3}\sin 3x + \frac{1}{5}\sin 5x + \ldots\right\}$

3. $f(x) = 1 + \frac{8}{\pi^2}\left\{\cos x + \frac{1}{3^2}\cos 3x + \frac{1}{5^2}\cos 5x + \ldots\right\}$

4. $f(x) = \frac{1}{2} + \frac{6}{\pi}\left\{\cos x - \frac{1}{3}\cos 3x + \frac{1}{5}\cos 5x - \ldots\right\}$

5. $f(x) = \frac{\pi}{2} + \left(\frac{2}{\pi} + 1\right)\cos x + \left(\frac{2}{3^2\pi} - \frac{1}{3}\right)\cos 3x + \left(\frac{2}{5^2\pi} + \frac{1}{5}\right)\cos 5x \ldots - \sin 2x - \frac{1}{3}\sin 6x - \ldots$

6. $f(x) = 4 - \frac{8}{\pi}\left\{\sin x + \frac{1}{3}\sin 3x + \frac{1}{5}\sin 5x + \ldots\right\}$

7. $f(x) = \frac{\pi^2}{3} + \frac{4}{\pi}\left\{-\cos x + \frac{1}{2}\cos 2x - \frac{1}{3}\cos 3x + \ldots\right\}$

8. $f(x) = \left(2 + \frac{4}{\pi}\right)\sin x - \sin 2x + \left(\frac{2}{3} - \frac{4}{3^2\pi}\right)\sin 3x - \frac{1}{2}\sin 4x + \left(\frac{2}{5} + \frac{4}{5^2\pi}\right)\sin 5x + \ldots$

9. $f(x) = \frac{5}{2} - \frac{20}{\pi^2}\left\{\sin x + \frac{1}{3^2}\sin 3x + \frac{1}{5^2}\sin 5x + \ldots\right\}$

10. $f(x) = 2 + 4\left\{\left(2 + \frac{1}{\pi}\right)\sin x - \frac{1}{2\pi}\sin 2x - \left(\frac{2}{3^2} - \frac{1}{3\pi}\right)\sin 3x - \frac{1}{4\pi}\sin 4x + \left(\frac{2}{5^2} + \frac{1}{5\pi}\right)\sin 5x + \ldots\right\}$

EXERCISE 10 (p. 42)

1. Even
2. Odd
3. Neither
4. Even
5. Neither
6. Even
7. Odd
8. Odd
9. Even
10. Neither

EXERCISE 11 (p. 45)

1. Even
2. Even
3. Odd
4. Odd
5. Neither
6. Even
7. Neither
8. Odd
9. Even
10. Odd

EXERCISE 12 (p. 49)

1. $f(x) = \frac{16}{\pi}\left\{\sin x + \frac{1}{3}\sin 3x + \frac{1}{5}\sin 5x + \ldots\right\}$

2. $f(x) = \frac{20}{\pi}\left\{\cos x - \frac{1}{3}\cos 3x + \frac{1}{5}\cos 5x - \ldots\right\}$

3. $f(x) = \frac{\pi^2}{3} + 2 - 4\left\{\cos x - \frac{1}{4}\cos 2x + \frac{1}{9}\cos 3x - \ldots\right\}$

4. $f(x) = \frac{12}{\pi}\left\{\sin x - \frac{1}{2}\sin 2x + \frac{1}{3}\sin 3x - \frac{1}{4}\sin 4x + \ldots\right\}$

5. $f(x) = \frac{2A}{\pi}\left\{2\sin x + \frac{1}{2}\sin 2x + \frac{4}{3}\sin 3x + \frac{3}{4}\sin 4x + \ldots\right\}$

6. $f(x) = -\left\{\sin x + \frac{1}{2}\sin 2x + \frac{1}{3}\sin 3x + \frac{1}{4}\sin 4x + \ldots\right\}$

EXERCISE 13 (p. 58)

1. $f(x) = \frac{4}{\pi}\left\{\sin x + \frac{1}{3}\sin 3x + \frac{1}{5}\sin 5x + \ldots\right\}$

2. $f(x) = \pi - \frac{8}{\pi}\left\{\cos x + \frac{1}{3^2}\cos 3x + \frac{1}{5^2}\cos 5x + \ldots\right\}$

3. $f(x) = \frac{6}{\pi}\left\{\sin x + \sin 2x + \frac{1}{3}\sin 3x + \frac{1}{5}\sin 5x + \ldots\right\}$

4. $f(x) = \frac{3\pi}{8} - \frac{2}{\pi}\left\{\cos x + \frac{1}{2}\cos 2x + \frac{1}{9}\cos 3x + \frac{1}{25}\cos 5x + \ldots\right\}$

5. $f(x) = \frac{4}{\pi}\left\{\cos x - \frac{1}{3}\cos 3x + \frac{1}{5}\cos 5x - \frac{1}{7}\cos 7x + \ldots\right\}$

6. $f(x) = \frac{2}{\pi}\left\{\sin x + \frac{\pi}{4}\sin 2x - \frac{1}{3^2}\sin 3x - \frac{\pi}{8}\sin 4x + \ldots\right\}$

7. $f(x) = \frac{16}{\pi}\left\{\sin 2x + \frac{1}{3}\sin 6x + \frac{1}{5}\sin 10x + \frac{1}{7}\sin 14x + \ldots\right\}$

8. $f(x) = \frac{1}{2} - \frac{2}{\pi}\left\{\cos x - \frac{1}{3}\cos 3x + \frac{1}{5}\cos 5x - \ldots\right\}$

9. $f(x) = \frac{\pi}{2} + 1 - \frac{4}{\pi}\left\{\cos x + \frac{1}{3^2}\cos 3x + \frac{1}{5^2}\cos 5x + \ldots\right\}$

10. $f(x) = \left(2\pi - \frac{8}{\pi}\right)\sin x - \pi\sin 2x + \left(\frac{2\pi}{3} - \frac{8}{\pi 3^3}\right)\sin 3x - \frac{\pi}{2}\sin 4x + \ldots$

EXERCISE 14 (p. 61)

1. Series contains both sine and cosine terms, but even harmonics only
2. Series contains both sine and cosine terms, but odd harmonics only
3. Series contains sine terms only and even harmonics only
4. Series contains cosine terms only and odd harmonics only
5. Series contains cosine terms only and even harmonics only
6. Series contains cosine terms only and odd harmonics only
7. Series contains cosine terms only and even harmonics only
8. Series contains cosine terms only and even harmonics only

9. Series contains sine terms only and even harmonics only
10. Series contains both sine and cosine terms, but odd harmonics only

EXERCISE 15 (p. 65)

1. $f(x) = 2 + \frac{16}{\pi}\left\{\cos x - \frac{1}{3}\cos 3x + \frac{1}{5}\cos 5x - \frac{1}{7}\cos 7x + \dots\right\}$
2. $f(x) = 8 + \frac{12}{\pi}\left\{\sin x + \frac{1}{3}\sin 3x + \frac{1}{5}\sin 5x + \dots\right\}$
3. $f(x) = 7 - \frac{6}{\pi}\left\{\sin x - \frac{1}{2}\sin 2x + \frac{1}{3}\sin 3x - \frac{1}{4}\sin 4x + \dots\right\}$
4. $f(x) = 2 + \frac{10}{\pi}\left\{\sin x - \frac{1}{2}\sin 2x + \frac{1}{3}\sin 3x - \frac{1}{4}\sin 4x + \dots\right\}$
5. $f(x) = 8 + \frac{16}{\pi^2}\left\{\cos x + \frac{1}{3^2}\cos 3x + \frac{1}{5^2}\cos 5x + \dots\right\}$
6. $f(x) = \frac{11}{2} - \frac{10}{\pi}\left\{\cos x - \frac{1}{3}\cos 3x + \frac{1}{5}\cos 5x - \dots\right\}$
7. $f(x) = -2 - \frac{24}{\pi}\left\{\sin x + \frac{1}{3}\sin 3x + \frac{1}{5}\sin 5x + \dots\right\}$
8. $f(x) = 5.5 - \frac{12}{\pi^2}\left\{\cos x + \frac{1}{3^2}\cos 3x + \frac{1}{5^2}\cos 5x + \dots\right\}$
9. $f(x) = 7 + \frac{6}{\pi}\left\{\sin x - \sin 2x + \frac{1}{3}\sin 3x + \frac{1}{5}\sin 5x + \dots\right\}$
10. $f(x) = 3 + \frac{8}{\pi^2}\left\{\sin x - \frac{1}{3^2}\sin 3x + \frac{1}{5^2}\sin 5x + \dots\right\}$
$+ \frac{4}{\pi}\left\{\sin x - \frac{1}{2}\sin 2x + \frac{1}{3}\sin 3x - \frac{1}{4}\sin 4x + \dots\right\}$

EXERCISE 16 (p. 69)

1. Put $u = \frac{\pi x}{4}$ i.e. substitute $x = \frac{4u}{\pi}$
2. Put $u = \frac{\pi x}{1}$ i.e. substitute $x = \frac{u}{\pi}$
3. Put $u = \frac{\pi x}{10}$ i.e. substitute $x = \frac{10u}{\pi}$
4. Put $u = \frac{\pi x}{2}$ i.e. substitute $x = \frac{2u}{\pi}$
5. Put $u = \frac{\pi x}{3}$ i.e. substitute $x = \frac{3u}{\pi}$

EXERCISE 17 (p. 74)

1. $f(x) = 3 + \frac{4}{\pi}\left\{\sin\frac{\pi x}{5} + \frac{1}{3}\sin\frac{3\pi x}{5} + \frac{1}{5}\sin \pi x + \frac{1}{7}\sin\frac{7\pi x}{5}\dots\right\}$
2. $f(x) = 5 - \frac{8}{\pi}\left\{\cos\frac{\pi x}{2} - \frac{1}{3}\cos\frac{3\pi x}{2} + \frac{1}{5}\cos\frac{5\pi x}{2} - \dots\right\}$
3. $f(x) = 6 - \frac{48}{\pi^2}\left\{\frac{1}{2^2}\cos\frac{2\pi x}{3} + \frac{1}{4^2}\cos\frac{4\pi x}{3} + \frac{1}{6^2}\cos\frac{6\pi x}{3} + \dots\right\}$

4. $f(x) = 4 + \frac{4}{\pi}\left\{\sin\frac{\pi x}{4} - \frac{1}{2}\sin\frac{\pi x}{2} + \frac{1}{3}\sin\frac{3\pi x}{4} + \ldots\right\}$

5. $f(x) = 1 + \frac{16}{\pi}\left\{\cos\frac{\pi x}{4} - \frac{1}{3}\cos\frac{3\pi x}{4} + \frac{1}{5}\cos\frac{5\pi x}{4} - \ldots\right\}$

6. $f(x) = \frac{50}{9} + \frac{100}{\pi^2}\left\{\cos\frac{\pi x}{5} - \frac{1}{2^2}\cos\frac{2\pi x}{5} + \frac{1}{3^2}\cos\frac{3\pi x}{5} - \ldots\right\}$

EXERCISE 18 (p. 79)

1. $f(t) = 2 + \frac{8}{\pi}\left\{\sin\omega t + \frac{1}{3}\sin 3\omega t + \frac{1}{5}\sin 5\omega t \ldots\right\}, \quad \omega = \frac{\pi}{4}$

2. $f(t) = 3 - \frac{36}{\pi^2}\left\{\cos\omega t - \frac{1}{2^2}\cos 2\omega t + \frac{1}{3^2}\cos 3\omega t \ldots\right\}, \quad \omega = \frac{\pi}{3}$

3. $f(t) = 1 - \frac{16}{\pi}\left\{\sin\omega t + \frac{1}{3}\sin 3\omega t + \frac{1}{5}\sin 5\omega t \ldots\right\}, \quad \omega = \frac{\pi}{2}$

4. $f(t) = \frac{8}{3} - \frac{16}{\pi^2}\left\{\cos\omega t + \frac{1}{2^2}\cos 2\omega t + \frac{1}{3^2}\cos 3\omega t \ldots\right\}, \quad \omega = \frac{\pi}{2}$

5. $f(t) = 2\left(\frac{1}{\pi} - \frac{6}{\pi^3}\right)\sin\omega t - \left(\frac{1}{2\pi} - \frac{6}{2^3\pi^3}\right)\sin 2\omega t + \ldots, \quad \omega = \pi$

6. $f(t) = 0.7 + 0.340\cos\omega t - 0.137\cos 2\omega t + 0.107\cos 3\omega t + \ldots$
$\quad - 0.208\sin\omega t + 0.029\sin 2\omega t + 0.042\sin 3\omega t + \ldots$ where $\omega = \frac{2\pi}{5}$

7. $f(t) = 1 - 1.17\cos\omega t + 0.328\cos 2\omega t + 0\cos 3\omega t + \ldots$
$\quad + 0.280\sin\omega t + 0.288\sin 2\omega t - 0.318\sin 3\omega t + \ldots$ where $\omega = \frac{\pi}{3}$

8. $f(t) = \frac{1 - e^{-4}}{2}\left\{\frac{1}{2} + \frac{1}{1+\omega^2}\cos\omega t + \frac{1}{1+4\omega^2}\cos 2\omega t + \ldots\right.$
$\left.\quad + \frac{\omega}{\omega^2 - 1}\sin\omega t + \frac{2\omega}{4\omega^2 - 1}\sin 2\omega t + \ldots\right\}$ where $\omega = \frac{\pi}{2}$

EXERCISE 19 (p. 99)

1. $f(x) = 1.16 + 0.275\cos x - 0.408\cos 2x + 0.167\cos 3x + \ldots$
$\quad + 0.671\sin x - 0.043\sin 2x + 0.050\sin 3x + \ldots$
$\quad = 1.16 + 0.725\sin(x + 0.389) - 0.410\sin(2x + 1.466)$
$\quad + 0.174\sin(3x + 1.280) + \ldots$

2. $f(x) = 5.30 - 0.133\cos x + 0.475\cos 2x - 0.667\cos 3x + \ldots$
$\quad + 8.472\sin x + 1.056\sin 2x - 0.717\sin 3x + \ldots$
$\quad = 5.30 + 8.47\sin(x - 0.0157) + 1.158\sin(2x + 0.423)$
$\quad - 0.979\sin(3x + 0.749) + \ldots$

3. $f(x) = 14.8 + 12.0\cos x - 0.333\cos 2x + 0.667\cos 3x + \ldots$
$\quad - 2.15\sin x - 3.460\sin 2x + 1.000\sin 3x + \ldots$
$\quad = 14.8 - 12.2\sin(x - 1.394) - 3.48\sin(2x + 0.0959)$
$\quad + 1.20\sin(3x + 0.5882) + \ldots$

4. $f(x) = 2.43 + 0.785 \cos x + 1.11 \cos 2x - 0.65 \cos 3x + \ldots$
$+ 11.1 \sin x - 0.332 \sin 2x + 1.17 \sin 3x + \ldots$
$= 2.43 + 11.1 \sin(x + 0.0706) - 1.16 \sin(2x - 1.280)$
$+ 1.34 \sin(3x - 0.5071) + \ldots$

5. $f(x) = 5.90 + 11.3 \cos x - 6.67 \cos 2x + 2.83 \cos 3x + \ldots$
$+ 18.8 \sin x - 8.08 \sin 2x - 1.00 \sin 3x + \ldots$
$= 5.90 + 21.9 \sin(x + 0.5412) - 10.5 \sin(2x + 0.6901)$
$- 3.00 \sin(3x - 1.2311) + \ldots$

6. $f(x) = 19.2 - 0.608 \cos x + 0.392 \cos 2x - 1.33 \cos 3x + \ldots$
$- 14.50 \sin x - 0.621 \sin 2x - 1.48 \sin 3x + \ldots$
$= 19.2 - 14.5 \sin(x + 0.0419) - 0.734 \sin(2x - 0.5631)$
$- 1.990 \sin(3x + 0.7321) + \ldots$

7. $f(x) = 2.64 + 1.21 \cos x - 0.795 \cos 2x - 0.317 \cos 3x + \ldots$
$+ 1.51 \sin x + 0.849 \sin 2x + 0.200 \sin 3x + \ldots$
$= 2.64 + 1.93 \sin(x + 0.6755) + 1.16 \sin(2x - 0.7526)$
$+ 0.375 \sin(3x - 1.008) + \ldots$

8. $f(x) = 27.1 \sin x - 0.667 \sin 3x + 2.25 \sin 5x + \ldots$

9. (a) $f(u) = 10.4 + 0.0578 \cos ku - 3.01 \cos 2ku - 1.03 \cos 3ku + \ldots$
$+ 6.27 \sin ku - 1.78 \sin 2ku + 1.73 \sin 3ku + \ldots$
$= 10.4 + 6.27 \sin(ku + 0.0092) - 3.50 \sin(2ku + 1.0367)$
$+ 2.01 \sin(3ku - 0.5370) + \ldots$ where $k = \frac{\pi}{30}$

(b) Second harmonic = 55.8%

10. $f(t) = 1.83 - 3.44 \cos \omega t + 1.83 \cos 2\omega t + 0.500 \cos 3\omega t + \ldots$
$+ 17.6 \sin \omega t - 6.06 \sin 2\omega t + 0.833 \sin 3\omega t + \ldots$
$= 1.83 + 17.9 \sin(\omega t - 0.1930) - 6.33 \sin(2\omega t - 0.2933)$
$+ 0.972 \sin(3\omega t + 0.5406) + \ldots$ where $\omega = \frac{\pi}{9}$

EXERCISE 20 (p. 115)

1. 3.861 V   2. 3.01 W   3. 1474 W

4.

n	1	2	3	4	5	6	7	8	9	10
A	3.07	0.50	1.00	0.25	0.60	0.17	0.43	0.12	0.33	0.10

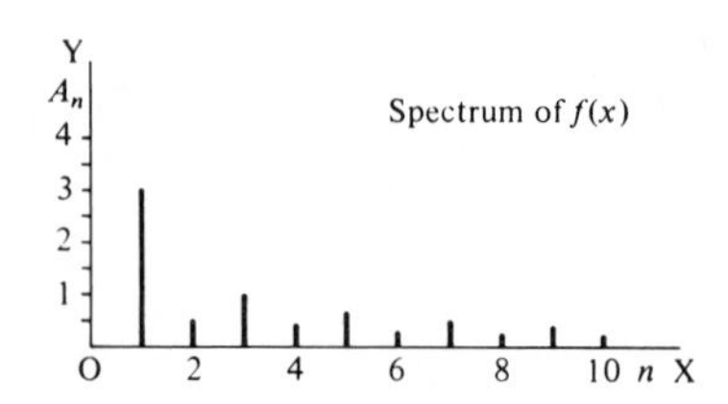

5.

n	1	2	3	4	5	6	7	8	9	10
A	0.398	0.140	0.115	0.077	0.064	0.055	0.044	0.040	0.035	0.032

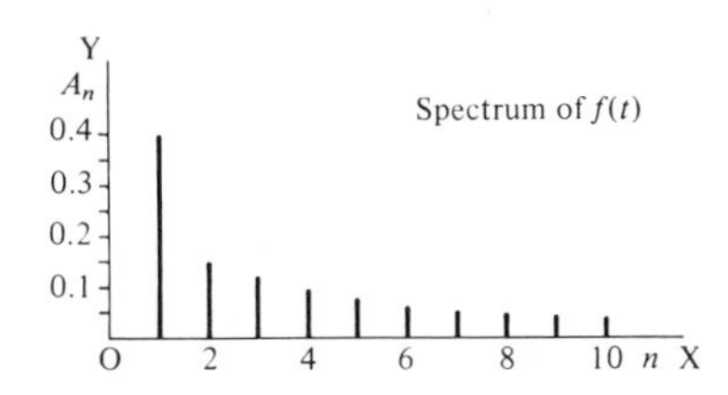

EXERCISE 21 (p. 125)

1. $j_1 = 2;\quad j_2 = -(3\pi + 2)$
2. $j_1 = -\pi^2$
3. $j_1 = -(\pi^2 - 4);\quad j_2 = -4$
4. $j_1 = -\left(3 + \dfrac{\pi^3}{80}\right) = -3.39;\quad j_2 = 3 + \dfrac{27\pi^3}{80} = 13.46$
5. $j_1 = -\dfrac{\pi}{2};\quad j_2 = \dfrac{\pi}{2}$
6. $j_1 = 8;\quad j_2 = -8$
7. $j_1 = -14;\quad j_2 = -2(\pi^2 - 7)$
8. $j_1 = -4;\quad j_2 = 2;\quad j_3 = -2$
9. $j_1 = -\pi;\quad j_2 = \pi$
10. $j_1 = -\left\{2 + \left(\dfrac{\pi}{3}\right)^2\right\} = -3.097;\quad j_2 = 2$

EXERCISE 22 (p. 133)

1. $$f(x) = \frac{10\pi^2}{3} + \frac{4}{\pi}\left\{\cos x + \frac{1}{2^3}\cos 2x + \frac{1}{3^3}\cos 3x + \ldots\right\} - 8\pi\left\{\sin x + \frac{1}{2}\sin 2x + \frac{1}{3}\sin 3x + \ldots\right\}$$

2. $$f(x) = \frac{\pi}{2} - 2 - \frac{4}{\pi}\left\{\cos x + \frac{1}{3^2}\cos 3x + \frac{1}{5^2}\cos 5x + \ldots\right\} + \frac{2}{\pi}\left\{(\pi + 4)\sin x - \frac{\pi}{2}\sin 2x + \frac{(\pi + 4)}{3}\sin 3x \ldots\right\}$$

3. $$f(x) = \frac{3\pi}{8} - \frac{2}{\pi}\left\{\cos x + \frac{1}{2}\cos 2x + \frac{1}{9}\cos 3x + \frac{1}{25}\cos 5x \ldots\right\}$$

4. $$f(x) = -\frac{1}{4} + \frac{10}{\pi^2}\left\{\cos x + \frac{1}{3^2}\cos 3x + \frac{1}{5^2}\cos 5x + \ldots\right\} - \frac{1}{\pi}\left\{15\sin x + \frac{1}{2}\sin 2x + \frac{15}{3}\sin 3x + \frac{1}{4}\sin 4x \ldots\right\}$$

5. $$f(x) = -\frac{4}{\pi}\left\{\left(1 + \frac{2}{\pi}\right)\sin x - \frac{1}{2}\sin 2x + \frac{1}{3}\left(1 - \frac{2}{3\pi}\right)\sin 3x - \frac{1}{4}\sin 4x + \frac{1}{5}\left(1 + \frac{2}{5\pi}\right)\sin 5x + \ldots\right\}$$

6. $$f(x) = \frac{\pi}{2} + \frac{1}{\pi}\left\{(\pi + 2)\cos x - \frac{1}{3}\left(\pi - \frac{2}{3}\right)\cos 3x + \frac{1}{5}\left(\pi + \frac{2}{5}\right)\cos 5x \ldots\right\} + \frac{1}{2}\sin 2x - \frac{1}{4}\sin 4x + \frac{1}{6}\sin 6x + \ldots$$

7. $f(x) = \frac{\pi^2}{6} - 1 - 2\left\{\cos x + \frac{1}{2^2}\cos 2x - \frac{1}{3^2}\cos 3x + \frac{1}{4^2}\cos 4x + \dots\right\}$
$+ \pi\left\{\sin x - \frac{1}{2}\sin 2x + \left(\frac{1}{3} + \frac{32}{3^3\pi^2}\right)\sin 3x - \frac{1}{4}\sin 4x \dots\right\}$

8. $f(x) = \frac{15\pi}{8} + \frac{2}{\pi}\left\{\cos x - \frac{1}{2}\cos 2x + \frac{1}{3^2}\cos 3x + \frac{1}{5^2}\cos 5x + \dots\right\}$
$+ \frac{1}{\pi}\left\{(3\pi + 2)\sin x - \frac{3\pi}{2}\sin 2x + \frac{1}{3}\left(3\pi - \frac{2}{3}\right)\sin 3x \dots\right\}$

9. $f(x) = \frac{1}{\pi}\left\{\left(\frac{3\pi^2}{2} - 16\right)\sin x + \frac{1}{2}\left(4 - \frac{3\pi}{2}\right)\sin 2x - \frac{1}{3}\left(\frac{32}{9} + \frac{\pi^2}{2}\right)\sin 3x \dots\right\}$

10. $f(x) = \frac{5\pi^2}{48} - \left\{\frac{\pi}{4}\cos x + \frac{1}{2}\cos 2x - \frac{\pi}{12}\cos 3x - \frac{1}{8}\cos 4x + \dots\right\}$
$+ \frac{\pi}{4}\left\{\sin x + \sin 2x + \frac{1}{3}\sin 3x + \frac{1}{5}\sin 5x + \dots\right\}$

EXERCISE 23 (p. 143)

1. $f(x) = \mathrm{j}\sum_{n=-\infty}^{\infty} \frac{(-1)^n}{n} \mathrm{e}^{\mathrm{j}nx}$

2. $f(x) = \sum_{n=-\infty}^{\infty} \frac{1 + \mathrm{j}n\pi}{1 + n^2\pi^2}\{\mathrm{e} - \mathrm{e}^{-1}\}(-1)^n \mathrm{e}^{\mathrm{j}n\pi x}$

3. $f(t) = \sum_{n=-\infty}^{\infty} \frac{1}{2}\{\mathrm{e}^{-2} - 1\}\frac{1 - \mathrm{j}n\omega}{1 + n^2\omega^2}\mathrm{e}^{\mathrm{j}n\omega t}$, where $\omega = \pi$

4. $f(x) = \sum_{n=-\infty}^{\infty} \frac{1}{2\pi}\left\{\frac{\mathrm{j}\pi}{n} - \frac{1 + (-1)^{n+1}}{n^2}\right\}\mathrm{e}^{\mathrm{j}nx}$

5. $f(x) = \sum_{n=-\infty}^{\infty} \left\{\frac{2}{n^2} + \mathrm{j}\frac{4\pi}{n}\right\}\mathrm{e}^{\mathrm{j}nx}$

6. $f(t) = \sum_{n=-\infty}^{\infty} \frac{A}{2\pi(n^2 - 1)}\{(-1)^n + 1\}\mathrm{e}^{\mathrm{j}n\omega t}$, where $\omega = \frac{2\pi}{T}$

7. $f(x) = \sum_{n=-\infty}^{\infty} \frac{1}{2\pi}\left\{\frac{1 + \mathrm{j}n}{1 + n^2}[\mathrm{e}^{\pi}(-1)^n - 1] - \mathrm{j}[(-1)^n - 1]\right\}\mathrm{e}^{\mathrm{j}nx}$

EXERCISE 24 (p. 165)

1. $u(x,t) = \frac{25}{2\pi^2}\sum_{r=1}^{\infty}\frac{1}{r^2}\sin\frac{r\pi}{5}\sin\frac{r\pi x}{10}\cos\frac{cr\pi t}{10}$ $\qquad r = 1, 2, 3, \dots$

2. $u(x,t) = \frac{32}{\pi^3}\sum_{r=1}^{\infty}\frac{1}{r^3}\sin\frac{r\pi x}{2}\cos\frac{3r\pi t}{2}$ $\qquad$ (r odd)

3. $u(x,t) = \frac{4}{\pi^2}\sum_{r=1}^{\infty}\frac{1}{r^2}\sin\frac{r\pi x}{20}\sin\frac{r\pi}{2}\cos\frac{cr\pi t}{20}$ $\qquad r = 1, 2, 3, \dots$

4. $u(x,t) = \frac{16}{\pi^2}\sum_{r=1}^{\infty}\frac{1}{r^2}\sin\frac{r\pi}{2}\sin\frac{r\pi x}{10}\cos\frac{r\pi t}{10}$ $\qquad r = 1, 2, 3, \ldots$

5. $u(x,t) = \frac{80}{\pi}\sum_{r=1}^{\infty}\frac{1}{r}\sin\frac{r\pi x}{5}\,e^{-r^2\pi^2c^2t/25}$ $\qquad r = 1, 3, 5, \ldots$

6. $u(x,t) = \frac{100}{\pi}\sum_{r=1}^{\infty}(-1)^{r+1}\frac{1}{r}\sin\frac{r\pi x}{l}\,e^{-r^2\pi^2c^2t/l^2}$ $\qquad r = 1, 2, 3, \ldots$

7. $u(x,t) = \frac{16}{\pi^2}\sum_{r=1}^{\infty}\frac{1}{r^2}\sin\frac{r\pi}{2}\sin\frac{r\pi x}{10}\,e^{-r^2\pi^2t/100}$ $\qquad r = 1, 3, 5, \ldots$

8. $u(x,y) = \frac{6}{\pi}\sum_{r=1}^{\infty}\frac{1}{r}\{1-(-1)^r\}e^{-ry}\sin rx$ $\qquad r = 1, 2, 3, \ldots$

9. $u(x,y) = \frac{20}{\pi}\sum_{r=1}^{\infty}\frac{1}{r}\sin\frac{r\pi x}{a}\sinh\frac{r\pi y}{a}$ $\qquad r = 1, 3, 5, \ldots$

10. $u(x,y) = -\frac{128}{\pi^3}\sum_{r=1}^{\infty}\frac{1}{r^3}\sin\frac{r\pi x}{4}\sinh\frac{r\pi}{4}(y-2)$ $\qquad r = 1, 3, 5, \ldots$

EXERCISE 25 (p. 177)

1. $f(x) = \frac{5}{\pi\lambda}\int_0^{\infty}\{\sin\lambda(1-x)+\sin\lambda x\}\,d\lambda$

2. $f(x) = \frac{2}{\pi}\int_0^{\infty}\cos\lambda x\left\{\frac{\cos\lambda-1}{\lambda^2}\right\}d\lambda$

3. $f(x) = \frac{2}{\pi}\int_0^{\infty}\sin\lambda x\left\{\frac{3\sin\lambda-\sin 3\lambda}{\lambda^2}\right\}d\lambda$

4. $$f(x) = \frac{1}{\pi}\int_0^{\infty}\left\{\left[\left(\frac{4}{\lambda}-\frac{2}{\lambda^3}\right)\sin 2\lambda+\frac{4}{\lambda^2}\cos 2\lambda\right]\cos\lambda x + \left[\left(\frac{2}{\lambda^3}-\frac{4}{\lambda}\right)\cos 2\lambda+\frac{4}{\lambda^2}\sin 2\lambda-\frac{2}{\lambda^3}\right]\sin\lambda x\right\}d\lambda$$

5. $f(x) = \frac{2}{\pi}\int_0^{\infty}(\cos\lambda-\cos 2\lambda)\cos\lambda x\,d\lambda$

EXERCISE 26 (p. 181)

1.

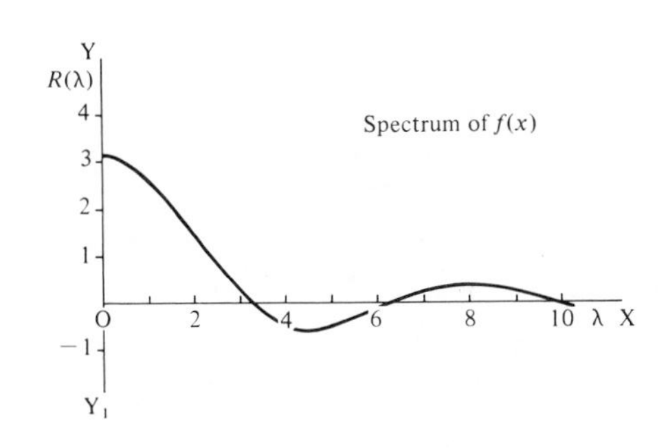

2.

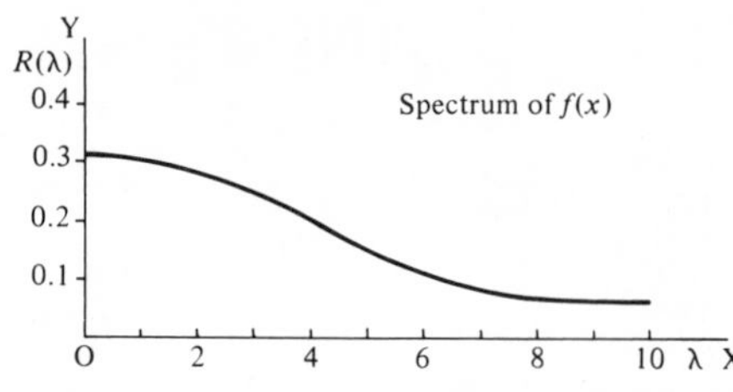

3.

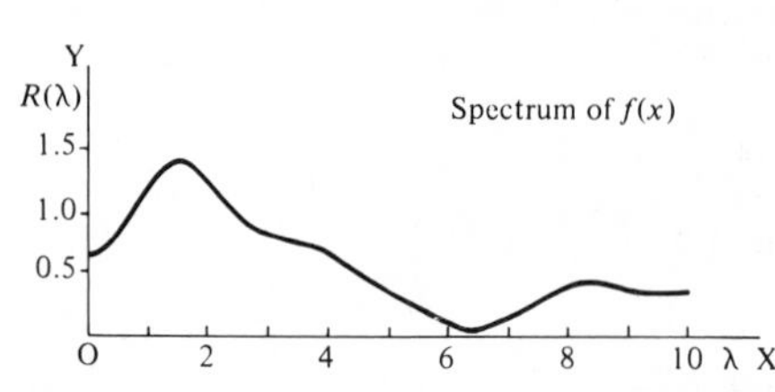

4.

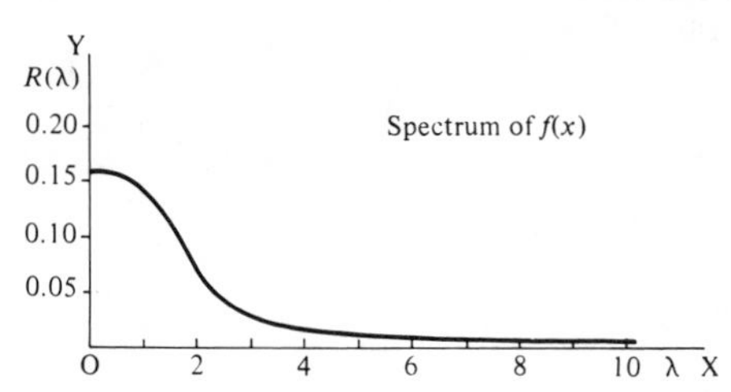

EXERCISE 27 (p. 184)

1. $F(\lambda) = \dfrac{4}{\lambda^3}\{\sin\lambda - \lambda\cos\lambda\}$

2. $F_s(\lambda) = \dfrac{1}{\lambda^2}\{\sin\lambda - \lambda\cos\lambda\}$

3. $F_c(\lambda) = \dfrac{2}{\lambda^2}\{1 - \cos\lambda + \lambda\sin 3\lambda\}$

4. $F_s(\lambda) = \dfrac{\lambda}{\lambda^2 + m^2}$

5. $F(\lambda) = \dfrac{j4}{\lambda^2 - 4}\sin\dfrac{\lambda\pi}{2}$

EXERCISE 28 (p. 189)

1. $v(x,t) = \dfrac{1}{\pi}\displaystyle\int_{-\infty}^{\infty} e^{-u}\int_0^{\infty} e^{-c^2\lambda^2 t}\cos\lambda(x-u)\,d\lambda\,du$

2. $v(x,t) = \dfrac{2}{\pi}\displaystyle\int_0^{\infty}\int_0^{\infty} f(u)\sin\lambda u\sin\lambda x\, e^{\lambda^2(1-t)}\,du\,d\lambda$

3. $v(x,y) = \dfrac{1}{\pi}\displaystyle\int_0^{\infty}\int_{-\infty}^{\infty} e^{-\lambda y} f(u)\cos\lambda(x-u)\,du\,d\lambda$

4. $v(x,t) = 2\displaystyle\sum_{r=1}^{\infty}\left\{\sin\frac{r\pi x}{2}\int_0^2\left(\frac{u^2}{r\pi c}\sin\frac{r\pi u}{2}\sin\frac{r\pi ct}{2} + u\sin\frac{r\pi u}{2}\cos\frac{r\pi ct}{2}\right)du\right\}$

5. $v(x,y) = \dfrac{2}{\pi}\displaystyle\int_0^{\infty}\frac{1}{\sinh 2\lambda}\left(\frac{\sin\lambda - \lambda\cos\lambda}{\lambda^2}\right)\sin\lambda x\sinh\lambda y\,dy$

INDEX

Amplitude 2, 178
Analysis, twelve-point 84
Analytical definitions 8
Applications of Fourier series 104
Approximate integration 83

Boundary conditions 147
Boundary value problems 146
and Fourier integrals 185

Change of units 67
Complex form of Fourier series 135, 137
Complex transform 181
Compound angle formulae 14
Compound waveforms 4
Conditions, Dirichlet 24
Constant term 24, 62
Convergence of Fourier series 39
Cosine integral 175
Cosine series, half-range 55, 76
Cosine transform 182
Cycle 1

Differential equations
ordinary 146
partial 147
Differentiation of Fourier series 120
Dirichlet conditions 24
Discontinuity, Fourier series 39

Eigenfunction 150
Eigenvalue 150
Equation, heat 156
Laplace 160
wave 147
Equivalent forms of Fourier series 142
Even functions 41
Even harmonics only 59
Extended period 169

Formulae, compound angle 14
sum and difference 14
Fourier coefficients 24, 27, 76, 135
from jumps 124, 127
Fourier complex transform 181
cosine integral 175
cosine transform 182
Fourier integrals 169, 170, 185
alternative forms 172
in boundary problems 185
Fourier series, applications 104
convergence 39
complex form 135, 137
differentiation 120
discontinuity 39
integration 119
even functions 45
odd functions 47
period 23
Fourier sine integral 174
Fourier sine transform 182
Fourier transforms 181
Full-wave rectifier 106
Functions, even 41, 45
odd 41, 47
of period 2π 23, 104, 106
of period $2L$ 67
of period T 75, 106, 108
Fundamental 76

Gibbs' phenomenon 121
Graphs of $A \sin n\theta$ 1

Half-range cosine series 55, 76
Half-range sine series 54, 77
Half-wave rectifier 104
Harmonics 2, 76
even only 4, 59
odd only 4, 59
Heat conduction equation 156
Hyperbolic functions 17
Hyperbolic identities 17

Identities, hyperbolic 17
trigonometrical 14
Integral, Fourier 169
Integral, Fourier cosine 175
Integral, Fourier sine 174
Integration, approximate 83
Integration by parts 18
Integration of Fourier series 119
Inverse transforms 181, 182, 183

Jumps, Fourier coefficients from 124
Jumps, positive and negative 124

Laplace equation 160

Multiplication theorem 111

Non-sinusoidal periodic functions 6
Numerical harmonic analysis 83

Odd and even functions, product 40, 43
Odd and even harmonics 4

Odd harmonics only 59
Odd functions 41
Ordinary differential equations 146

Parseval's theorem 108
Partial differential equations 147
Period 1
Period of 2π 23
Period of $2L$ 67
Period T 75
Period, extended 169
Periodic functions 1, 23
  analytical definition 8
  sketching graphs 11
Periods other than 2π 67
Phase 178
Power in a circuit 112
Product of odd and even functions 43

Rectifier, half-wave 104
  full-wave 106
Root mean square value 109

Separating the variables 148
Sine integral 174
Sine series, half-range 54, 77
Sine transform 182
Sketching periodic functions 11
Spectrum of a waveform 113, 180
Sum and difference formulae 14
Symmetry about origin 42
  about the y-axis 41

Transform, Fourier complex 181
  Fourier cosine 182
  Fourier sine 182
  inverse 181, 182
Trapezoidal rule 83
Trigonometrical identities 14
Twelve-point analysis 84
  tabular form 87

Useful integrals 26
Useful revision 14

Wave equation 147
Waveform, spectrum of 113, 180